CAN YOU
SOLVE MY
PROBLEMS?

Alex Bellos is brilliant on all things mathematical. He has a degree in Mathematics and Philosophy from Oxford University. His bestselling books, *Alex's Adventures in Numberland* and *Alex Through the Looking-Glass*, have been translated into over 20 languages and were both short-listed for the Royal Society Science Book prize. He is the co-author of two mathematical colouring books and he has launched an elliptical pool table, LOOP. He writes a popular maths blog and a puzzle blog for the *Guardian*. He won the Association of British Science Writers award for best science blog in 2016.

'Think of the best storyteller you know and the coolest teacher you ever had, and now you've got some idea of what Alex Bellos is like.' Steven Strogatz, Cornell University

'Superbly engrossing.' *Spectator*

CAN YOU SOLVE MY PROBLEMS?

A casebook of ingenious, perplexing
and totally satisfying puzzles

Alex Bellos

gff

First published in 2016
by Guardian Books, Kings Place, 90 York Way, London, N1 9GU
and Faber & Faber Limited
Bloomsbury House, 74–77 Great Russell Street
London WC1B 3DA
This paperback edition published in 2017

A CIP record for this book is available from the British Library

ISBN 978-1-78335-1152

Typeset by carrdesignstudio.com
Diagrams by ruthmurray.net
Cartoons by www.doodlesandstuff.com

Printed in the UK by CPI Group (UK) Ltd, Croydon, CR0 4YY

4 6 8 10 9 7 5 3

To Zak

CONTENTS

INTRODUCTION

All my problems began with Cheryl.

She was a complicated girl. A real tease.

But I couldn't stop thinking about her. And in many ways she changed the course of my life.

I should clarify here that Cheryl doesn't exist. She was the protagonist in a question from a Singaporean maths exam that captured my imagination, seducing me into exploring a world of puzzles that led to this book.

You will find the Cheryl's Birthday problem – and the full story of our relationship – later in this book. (It's Problem 21, and appears on page 41.) But before we embark on our journey through my favourite mathematical puzzles, here are two appetite-whetting *amuse-brains*.

First, look at the image below. The numbers are arranged according to a certain rule. Once you've worked out the rule, fill in the missing number. The number seven in the final circle is not a typographical error.

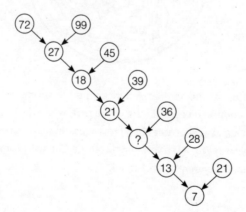

I find this puzzle irresistible. It looks intriguing. It requires no advanced mathematics. It taunts you to solve it, and when you do (if you do) the sense of satisfaction is both exhilarating and addictive. Nob Yoshigahara, a celebrated twentieth-century Japanese puzzle inventor, considered it his masterpiece. Try to work out the answer before I reveal it, at the end of this section.

Second, the Canals on Mars. A map of the red planet displays newly discovered cities and waterways. Start at the city T on the south pole. Travelling along the canals, visiting each city only once and returning to the starting point, can you spell out a sentence in English?

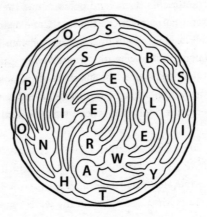

This problem, by the prolific American puzzle inventor Sam Loyd, is more than a hundred years old. He wrote that 'when the puzzle originally appeared in a magazine, more than fifty thousand readers reported "There is no possible way." Yet it is a very simple puzzle.' You will kick yourself if you read the answer before you work it out.

* * *

If you paused to tackle either of these two problems I hardly need explain why puzzles are so enjoyable. They reel you in. Distractions around you disappear as you focus in on the solution. Being forced to use your wits is life-affirming. And deductive reasoning in simple logical steps is comforting, especially when real life is often so illogical. Good puzzles also present achievable goals, which are supremely gratifying when they come.

One consequence of my tryst with Cheryl was that I began to write an online puzzle column for the *Guardian*. In order to find the best possible puzzles I immersed myself in books and corresponded with both amateur and professional puzzlesmiths. I have always loved mathematical puzzles, but before I began this research I had not fully appreciated their variety, conceptual depth and rich history. I did not realise, for example, that one thousand years ago the main role of mathematics – with the exception of dull commercial tasks like counting and measuring – was to provide intellectual diversion and amusement. (Arguably, the statement remains true today, given that the number of Sudoku enthusiasts vastly exceeds the number of professional mathematicians.) Puzzles form a parallel history of mathematics, reflecting great discoveries and inspiring the sharpest minds.

This book is a curated collection of 125 brainteasers from the last two millennia, linked with stories about their origins and influence. I have chosen the puzzles that I found most fascinating, entertaining and thought-provoking. They are mathematical only in the widest possible sense: the solutions require logical thought, but they do not require advanced maths. The problems come from ancient China, medieval Europe, Victorian England and modern-day Japan, as well as other ages and places. Some are traditional riddles, others were devised by the top professional mathematicians of their day. Often, however, it's hard to say where a puzzle comes from. Like jokes and folk tales, they are always evolving as each new generation embellishes, adapts, simplifies, extends and restyles them.

The best puzzles are pieces of poetry. With elegance and brevity, they pique our interest, kindle our competitive spirit, test our ingenuity, and in some cases reveal universal truths. A good puzzle requires no specialist knowledge – only creativity, cunning and an ability to think clearly. Puzzles are captivating because they appeal to the human impulse to make sense of the world; they give us pleasure because we are making sense out of something. Yet no matter how frivolous or contrived a puzzle may be, the strategies we use to solve them expand our armoury for tackling other challenges in life.

Most importantly, however, puzzles indulge our intellectual playfulness. They are fun. They reflect a childlike sense of curiosity. I've chosen puzzles in as many styles as possible, and which require us to think in very different ways. Some rely on a flash of insight, some require us to follow our noses, and others – well, that would be telling.

Each chapter has a theme, and its problems appear roughly in chronological order. The questions are *not* in order of difficulty. Level is often hard to judge, anyway. What's torture for one person may be trivial for another, and vice versa. I explain how to solve a few puzzles, and give tips for a few more, but for the rest you're on your own. (The answers are at the back of the book.) Some of my problems are simple. Some will have you scratching your head for days. These toughies are signposted with the symbol ☞. If you fail to crack them I hope you will find the solutions just as fascinating as the problems themselves. Sometimes the thrill is in learning a technique, an idea, or its consequences, that you did not know.

Before each chapter I have included ten quick-fire questions to get you in the mood. The first, third and fifth contain questions, of increasing difficulty, used by the United Kingdom Mathematics Trust in their annual national challenges for students aged 11 to 13. That's right, ten questions *for children*. Are you up to it?

Now back to those questions I asked at the start.

When you look at the 'number tree' your eye goes to the top left. How can 72 and 99 make 27?

Got it! 99 – 72 = 27.

In other words, the number in a circle is the difference of the numbers in the two circles that point to it.

And look, the same pattern follows with 18, the next number down the line:

45 – 27 = 18

And with 21 too:

39 – 18 = 21

Which means that the missing number must be the difference between 21 and 36, which is 15:

36 – 21 = 15

For completeness's sake, we continue down the tree:

28 – 15 = 13

Nice! So it continues to work. We're almost there.

Until BOOM.

The very last number is 7, which is *not* the difference between 13 and 21, the two numbers pointing to it.

Dammit! Our original supposition is wrong. It is *not* the case that a number in a circle is equal to the difference of the two numbers pointing to it. Yoshigahara has artfully led us down the garden path, only to trip us up at the end of it.

Back to square one. Or rather circle one. How else can 72 and 99 produce 27?

The answer is so simple you may have missed it.

7 + 2 + 9 + 9 = 27

The digits all add up.

And this works for the next line:

2 + 7 + 4 + 5 = 18

And the next. So the missing number must be:

$2 + 1 + 3 + 6 = 12$

The final two circles now fall in line:

$1 + 2 + 2 + 8 = 13$

$1 + 3 + 2 + 1 = 7$

The puzzle is fantastically ingenious because Yoshigahara has found two arithmetical rules that fit the same numbers for *five* steps in the sequence, but only one of which fails for the final step, and then only by 1. The effortlessness in which it sets us off in the wrong direction is magical. Often a problem is difficult not because it is a 'difficult problem' but because we are going about it the wrong way. Take note.

Did you solve the Canals on Mars puzzle? You can trace the sentence 'There is no possible way'. The lesson here is *read the words carefully*.

Let's get puzzling.

Ten tasty teasers

Are you smarter than an 11-year-old?

Class rules: no calculators allowed.

1) The diagrams on the right show three different views of the same cube. Which letter is on the face opposite U?

A I **B** P **C** K **D** M **E** O

2) Pinocchio's nose is 5cm long. Each time he tells a lie his nose doubles in length.
After he has told nine lies, his nose will be roughly the same length as a:

A domino **B** tennis racket **C** snooker table **D** tennis court
E football pitch

3) The word 'thirty' contains 6 letters, and $30 = 6 \times 5$. Similarly, the word 'forty' contains 5 letters, and $40 = 5 \times 8$.
Which of the following is not a multiple of the number of letters it contains?

A six **B** twelve **C** eighteen **D** seventy **E** ninety

4) Amy, Ben and Chris are standing in a row.
If Amy is to the left of Ben and Chris is to the right of Amy, which of these statements must be true?

A Ben is furthest to the left   **B** Chris is furthest to the right
C Amy is in the middle   **D** Amy is furthest to the left
E None of the statements A, B, C, D is true.

5) Which of these diagrams can be drawn without taking the pen off the page, and without drawing along a line that's already been drawn?

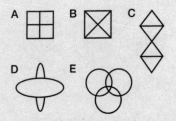

6) What is the remainder when 354972 is divided by 7?

A 1 **B** 2 **C** 3 **D** 4 **E** 5

7) Among the children in a certain family, each child has at least one brother and at least one sister.
What is the *smallest* possible number of children in the family?

A 2 **B** 3 **C** 4 **D** 5 **E** 6

8) The number 987654321 is multiplied by 9.
How many times does the digit 8 appear in the result?

A 1 **B** 2 **C** 3 **D** 4 **E** 9

9) In this partially completed pyramid, each rectangle is to be filled with the sum of the two numbers in the rectangles immediately below it.
What number should replace x?

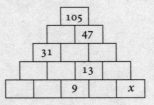

A 3 **B** 4 **C** 5 **D** 7 **E** 12

10) How many different digits appear when $\frac{20}{11}$ is written as a recurring decimal?

A 2 **B** 3 **C** 4 **D** 5 **E** 6

Cabbages, Cheating Husbands and a Zebra

LOGIC PROBLEMS

Logic. It's a logical place to start: logical deduction is the ground rule of all mathematical puzzles. Indeed, logic is the foundation of all mathematics. In the nomenclature of puzzledom, however, 'logic problems' are brainteasers that employ deductive reasoning alone – shunning, for example, any type of arithmetical calculation, algebraic manipulation, or sketching of shapes on the backs of envelopes. They are the most accessible type of mathematical conundrum because they require no technical knowledge, and the questions easily lend themselves to humorous phrasing. But, as we shall see, they are not always the easiest to solve, since they twist our brains in unfamiliar ways.

Which they have been doing since at least the time of Charlemagne, King of the Franks.

In 799 CE, Charlemagne, who ruled over much of Western Europe, received a letter from his old teacher, Alcuin: 'I have sent you', it read, 'some arithmetical curiosities to amuse you.'

Alcuin was the greatest scholar of his era. He grew up in York, attending and then running the city's cathedral school, the best educational establishment in the country. The Englishman's reputation reached Charlemagne. The king persuaded him to run his palace school in Aachen, where Alcuin founded a large library and went on to reform education across the Carolingian empire. Alcuin eventually left Charlemagne's court to become Abbot of Tours, which is when he wrote the above letter to his former boss.

Alcuin is credited by some with inventing joined-up writing so he and his many scribes could write faster. Some believe that he was also the first person to use a symbol – a diagonal squiggle – as punctuation for a question. It is wonderfully appropriate that the predominant early figure in the history of puzzles was also a progenitor of the question mark.

The physical document that Alcuin was referring to in his letter to Charlemagne no longer exists, but historians believe that it was a list of fifty or so problems called *Propositiones ad Acuendos Juvenes*, or *Problems to Sharpen the Young*, of which the earliest surviving manuscript dates from a century later. Who else, they argue, could have written it but Alcuin, the foremost teacher of his day?

Propositiones is a remarkable document. It is the largest cache of puzzles from medieval times, as well as the first Latin text that contains original mathematical material. (The Romans may have built roads, aqueducts, public baths and sanitation systems, but they never did any mathematics.) It begins in thigh-slapping tone:

> *A swallow invites a snail for lunch a league away. If the snail travels an inch a day, how long will it take him?*

The answer is 246 years and 210 days. He would have died more than two centuries before he got there.

Another one asks:

> *A certain man met some students and asked them: 'How many of you are there in your school?' One of the students replied: 'I don't want to tell you directly but I'll tell you how to work it out. You double the number of students, then triple that number, then divide that number into four parts. If you add me to one of the quarters, there will be 100.' How many students are there in the school?*

Pesky kids! I'll leave this one for you to solve on your own.

Alcuin's whimsical phrasing was groundbreaking. It was the first time humour had been used to pique students' interest in arithmetic. Yet *Propositiones* was important not just because of its stylistic innovations, but also because it included several new types of problem. Some of them required deductive reasoning but no calculation. The best known of Alcuin's puzzles is arguably the most famous mathematical riddle of all time.

(1)

WOLF, GOAT AND CABBAGES

A man arrives at a riverbank with a wolf, a goat and a bunch of cabbages. He needs to cross the river, but the one boat available can carry only him and a single item at the same time. He cannot leave the wolf alone with the goat or the goat alone with the cabbages, since in both cases the former will eat the latter.

How does he cross the river in the shortest number of crossings?

It's a blinder for two reasons. For a start, the scene is comical. You've spent all morning trudging down a dirt path, desperately trying to keep the wolf away from the goat, and the goat away from the cabbages. Now your day just got worse: you have to cross a river in a stupidly tiny boat. Yet what I find most amusing and interesting about the scenario is its solution, which forces our hero to behave in a way you would not intuitively expect.

Have a go. All five-year-old children can solve it, declared one thirteenth-century text. No pressure.

Or follow the reasoning with me.

Let's place the traveller on the left bank. He begins with three items and can only take one of them in the boat. If he takes the wolf, the goat will be left with the cabbages and eat them. If he takes the cabbages, the wolf will eat the goat. By a process of elimination, the only item he can take on the first crossing is the goat, since wolves don't eat cabbages. He delivers the goat to the right bank and returns for the next item.

Now he has a choice of wolf or cabbages. Let's say he decides to take the cabbages. He crosses the river for the third time. When he reaches the right bank he cannot leave the cabbages with the goat, so what does he do? He makes no progress if he returns with the cabbages, since he only just arrived with them, so he must return with the goat. This step is the counter-intuitive one: in order for the traveller to get everything across he needs to take something across, back, and across again.

Back on the left bank, after four crossings, with the wolf and the goat, the traveller chains the goat and departs for his fifth traverse with the wolf. On the right bank the wolf remains uninterested in the cabbages. All that is left is one trip back to pick up the bearded bovid, and our chap is done, in seven crossings.

(There's a second, equivalent, solution: if he took the wolf on the second crossing, the same logic follows and he also finishes the job in seven trips.)

Propositiones also contains other river-crossing puzzles, such as this one, which sounds like the plot of a bedroom farce.

$$2$$

THREE FRIENDS AND THEIR SISTERS

So there were three of us men who each had a sister, and we all had to get across a river. But each of us lusted after one of the others' sisters. On coming to the river all we found was a little ferry boat that could only carry two people at a time. Say if you can how we crossed the river so that none of our sisters was dishonoured by being alone in a boat with a man who was not her brother.

You can interpret this problem in two ways, since Alcuin's phrasing is ambiguous. What's not in dispute is that there are three pairs, each consisting of a brother and a sister, who must all cross the river, and all that they have at their disposal is a two-person boat. But there could be either of two restrictions: [1] That a boat can never contain a woman and a man who are not related. In this case the entire party can reach the other side in nine crossings. [2] That a woman is forbidden from being in the boat unaccompanied by her brother when the boat is dropping off or collecting passengers at a bank where there are other men. The second scenario, I think, is more in the spirit of the question, and the mission requires eleven crossings. Try to find both solutions.

River-crossing puzzles have delighted children and adults for more than a thousand years. As they have spread across the world, they have changed to reflect local concerns. In Algeria, the wolf, goat and cabbages are a jackal, a goat and a bundle of hay; in Liberia they are a cheetah, a fowl and some rice; and in Zanzibar they are a leopard, a goat and some leaves. The puzzle of the three friends and their sisters has also evolved throughout the ages: the lecherous men soon became jealous husbands forbidding their wives to travel in the boat with another man. In one thirteenth-century retelling the couples have names: Bertoldus and Berta, Gherardus and Greta, and Rolandus and Rosa. The solution is presented as two hexameters. If you can read Latin, look away now:

> *Binae, sola, duae, mulier, duo, vir mulierque,*
> *Bini, sola, duae, solus, vir cum muliere.*

By the seventeenth century the couples were masters and valets. Each master forbade his valet to travel with another master in case that master murdered him. The social warfare was reversed in the nineteenth century: the couples became masters and servants, and servants were forbidden to outnumber masters on any one side in case they were tempted to rob their bosses. Then xenophobia replaced the sexism and class war: the classic version became a travelling party of three missionaries and three hungry cannibals. You learn as much about the evolution of social stereotypes from this puzzle as you do about mathematics.

The following river-crossing puzzle emerged in the 1980s. By the turn of the century it was being used by Microsoft as one of its notoriously tricky interview questions to test the problem-solving skills of prospective employees. The key here is to let your logical brain overrule your gut instinct.

(3)

CROSSING THE BRIDGE
(WITH A LITTLE HELP FROM MY FRIENDS)

Four people – John, Paul, George and Ringo – are at one side of a gorge connected to the other by a rickety bridge that can only carry two people at a time. It is night time, and the structure is precarious, so whoever is crossing must use a torch. The group has a single torch, and the gorge is too wide for them to be able to throw it from one side to the other, so the torch must be walked back and forth over the bridge as the people cross. John can cross the bridge in 1 minute, Paul in 2, George in 5 and Ringo in 10. If two people cross together, they walk at the speed of the slowest of the two.

How does the foursome get over in the quickest possible time?

The obvious way to solve this problem is for John to accompany each friend one by one, since he can return fastest to pick up the next person. This strategy gets everyone over in 2 + 1 + 5 + 1 + 10 = 19 minutes. But is there a faster way?

Back to Alcuin, and a question from *Propositiones*:

> How many footprints are left in the last furrow by an ox which has been ploughing all day?

None, of course! The plough will have destroyed them all. It's the earliest trick question in puzzle literature.

Among the other types of puzzle that debuted in *Propositiones* is the 'kinship riddle', in which you have to find the relationships in unconventional families. It's my final pick from the old Yorker, before we fast-forward a thousand years.

(4)

THE DOUBLE DATE

If two men each take the other's mother in marriage, what would be the relationship between their sons?

I find kinship riddles hugely entertaining. No matter how straight-faced and logical I try to remain when solving them, I cannot help but speculate about the bizarro-weird backstories.

This type of puzzle has been a staple since medieval times and was much enjoyed by the Victorians, who perhaps found the suggestion of subversion in traditional family structures especially titillating.

Lewis Carroll was a fan. The next problem is taken from one of the chapters – or 'knots', as he called them – of his 1885 book *A Tangled Tale*. I consider it the apogee of the genre.

(5)

THE DINNER PARTY

The Governor of – what-you-may-call-it – wants to give a *very* small dinner party, and he means to ask his father's brother-in-law, his brother's father-in-law, his father-in-law's brother, and his brother-in-law's father: and we're to guess how many guests there will be.

How many guests are there, if the dinner party is to be as small as possible?

Through his novels *Alice's Adventures in Wonderland* and *Through the Looking-Glass*, Carroll is probably the writer who has done most to popularise the fun to be had with logic. Both stories are full of paradoxes, games and philosophical riddles. Carroll, the pen name of Charles Lutwidge Dodgson, a maths don at Oxford, also wrote three books of mathematical puzzles. None had anything like the success of the Alice books, partly because the maths was too difficult.

Carroll was, however, one of the first to devise puzzles based on truth-tellers and liars, a type of logical conundrum that would later become popular. He noticed that if different people are accusing each other of being liars then it may be possible to deduce who is telling the truth. 'I have worked out in the last few days some curious problems on the plan of "lying" dilemma', he wrote in his diary in 1894, mentioning the following puzzle, rephrased here with familiar characters. The puzzle was printed as an anonymous pamphlet later that year.

(6)

LIARS, LIARS

Berta says that Greta tells lies.
Greta says that Rosa tells lies.
Rosa says that both Berta and Greta tell lies.
Who is telling the truth?

We'll return to truth-tellers and liars shortly.

But before we get there, can you solve the logic puzzle that went viral in the early 1930s?

(7)

SMITH, JONES AND ROBINSON

Smith, Jones and Robinson are the driver, fireman and guard on a train, but not necessarily in that order. The train carries three passengers, coincidentally with the same surnames, but identified with a 'Mr': Mr Jones, Mr Smith and Mr Robinson.

Mr Robinson lives in Leeds.

The guard lives halfway between Leeds and Sheffield.

Mr Jones's salary is £1,000 2s. 1d. per annum.

Smith can beat the fireman at billiards.

The guard's nearest neighbour (one of the passengers) earns exactly three times as much as the guard.

The guard's namesake lives in Sheffield.

What is the name of the engine driver?

(I have kept the original phrasing of the puzzle, which uses the old British currency. The importance of the value £1,000 2s. 1d., or one thousand pounds, two shillings and one pence, is that you cannot divide it by three to produce an exact amount.)

I love this puzzle. It invites you to become a detective. On first reading it looks like there is far too little information to find the answer. But slowly, piecing together the clues, you will uncover the correct identities.

Not long after Smith, Jones and Robinson appeared in *The Strand Magazine*, a London literary journal, in April 1930, it became a British craze, reprinted in newspapers up and down the country. It spread around the world, and by 1932 the *New York Times* was reporting on the puzzle's popularity, presenting an Americanised version, with Leeds and Sheffield swapped for Detroit and Chicago.

The most straightforward way to solve the puzzle is to draw two grids. I'll start you off. We need to find which of Smith, Jones and Robinson are the driver, fireman and guard, so draw one grid, as shown below left, that contains the names of the workers and the professions. The question also involves three passengers and three locations. So draw a second grid, shown below right, that has Mr Smith, Mr Jones and Mr Robinson, against Leeds, Sheffield and halfway between.

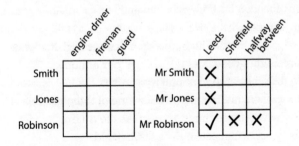

Our first piece of solid information is that Mr Robinson lives in Leeds, so we can tick the Mr Robinson/Leeds box, and put crosses in the boxes that have Mr Robinson living somewhere else, or anyone else living in Leeds. We need to take other clues together to fill in more squares. For example, the guard's nearest neighbour, who is a passenger, earns exactly three times as much as the guard does. So we can eliminate Mr Jones as the guard's nearest neighbour, because his salary is indivisible by three. I'll leave the final sleuthing to you.

The creator of Smith, Jones and Robinson died the month it appeared. Henry Ernest Dudeney was seventy-three and had been writing puzzles for *Strand* for more than twenty years. He was the most brilliant mathematical puzzle designer of his age, and yet it was only posthumously, with Smith, Jones and Robinson, that he had arguably his greatest success. When the *New Statesman* republished the puzzle, 'the result was astonishing,' wrote Hubert Phillips, editor of the magazine's bridge column and crossword. 'A flood of solutions (though none had been asked for) showed how wide and interested a public there is for inferential puzzles.'

Phillips himself was a former economics lecturer and adviser to the Liberal Party; in his early forties at the time of the puzzle, he had recently moved into journalism. Unprecedented interest in the puzzle caused Phillips to abandon his bridge column and replace it with a regular logic problem. Through the 1930s Phillips became a prolific and innovative creator of mathematical (and other) puzzles, turning the decade into a golden age for the genre.

I really like these two next problems of his. The first is a whodunnit. Or maybe a shedunnit. The second is a witty nod to the tradition of kinship riddles.

(8)

ST DUNDERHEAD'S

St. Dunderhead's School at Fogwell has a high reputation for
hockey – but not so high a reputation for veracity. The First
XI played a match at Diddleham recently, after which the
girls were allowed to go to a concert. Miss Pry, the mistress
in charge, collected the team afterwards; she saw ten girls
emerge from the concert hall and one from the cinema next
door. When she asked who had been to the cinema, the
members of the team replied as follows:

Joan Juggins:	'It was Joan Twigg.'
Gertie Gass:	'It was I.'
Bessie Blunt:	'Gertie Gass is a liar.'
Sally Sharp:	'Gertie Gass is a liar, and so is Joan Juggins.'
Mary Smith:	'It was Bessie Blunt.'
Dorothy Smith:	'It was neither Bessie nor I.'
Kitty Smith:	'It wasn't any of us Smith girls.'
Joan Twigg:	'It was either Bessie Blunt or Sally Sharp.'
Joan Forsyte:	'Both of the other Joans are telling lies.'
Laura Lamb:	'Only one of the Smith girls is telling the truth.'
Flora Flummery:	'No, two of the Smith girls are telling the truth.'

Given that, of these eleven assertions, at least seven are
untrue, who went to the cinema?

(9)

A CASE OF KINSHIP

There must have been a dearth of eligible young ladies in Kinsleydale, for each of five men there has married the widowed mother of one of the others. Jenkins's stepson, Tomkins, is the stepfather of Perkins. Jenkins's mother is a friend of Mrs Watkins, whose husband's mother is a cousin of Mrs Perkins.

What is the name of the stepson of Simkins?

Logic problems like the ones above are now commonly known as 'grid' puzzles, because the best way to solve them is to draw a grid showing all possible options. The most famous of the genre, *The Zebra Puzzle*, dates from the 1960s and is of unknown authorship.

The puzzle first appeared in *Life International* magazine in 1962. Often it is called Einstein's Riddle, since apparently he wrote it, although that would be pretty impressive given that he died in 1955. Another claim frequently made about the puzzle is that only 2 per cent of the population can solve it. This is probably untrue, but it is a brilliant tease.

(10)

THE ZEBRA PUZZLE

1. There are five houses.
2. The Scot lives in the red house.
3. The Greek owns the dog.
4. Coffee is drunk in the green house.
5. The Bolivian drinks tea.
6. The green house is immediately to the right of the ivory house.
7. The brogue wearer owns snails.
8. Brothel creepers are worn in the yellow house.
9. Milk is drunk in the middle house.
10. The Dane lives in the first house.
11. The person who wears Birkenstocks lives in the house next to the person with the fox.
12. Brothel creepers are worn in the house next to the house where the horse is kept.
13. The slipper wearer drinks orange juice.
14. The Japanese wears Havaianas.
15. The Dane lives next to the blue house.

Now, who drinks water? Who owns the zebra?

For clarification, each of the five houses is painted a different colour, and their inhabitants are of different national extractions, own different pets, drink different beverages and wear different types of shoes. In *Life's* version, the neighbours smoked different brands of American cigarettes. I have replaced this category with footwear since Einstein was famous for never wearing socks.

The response from *Life*'s readers was overwhelming. 'The magazine had barely gone on sale when the letters began to flood our mail room,' the editor wrote in a following edition, which featured the puzzle on the cover. 'They came from lawyers, diplomats, doctors, engineers, teachers, physicists, mathematicians, colonels, privates, priests and housewives – and from some astonishingly learned and logical children. The writers lived thousands of miles apart – in the provincial villages of England, in the Faroe Islands, in the Libyan Desert, in New Zealand – but they enjoyed one gift in common, an extraordinarily high level of intelligence.' Reader, don't let me down.

If you enjoyed that, you will appreciate the brain-mangling brilliance of the next puzzle. Devised by Max Newman, a young Cambridge logician, it appeared in Hubert Phillips's *New Statesman* column in 1933. Phillips used the pseudonym Caliban for his *New Statesman* puzzles, after the enslaved demi-devil of Shakespeare's *The Tempest*. Many of his Caliban puzzles were collaborations with professional mathematicians, and this one was by far the most wicked.

The puzzle is a work of genius. The information presented in the question seems comically insufficient for a solution but, of course, it contains just what you need. *The Mathematical Gazette* said Newman's puzzle was a 'gem' and 'must be solved to be believed.' I struggled with this one too, but that didn't stop me marvelling at its sparseness, and whimpering at the brutal elegance of its solution.

11

CALIBAN'S WILL

When Caliban's will was opened it was found to contain the following clause:

'I leave ten of my books to each of Low, Y.Y. and "Critic", who are to choose in a certain order:

[1] No person who has seen me in a green tie is to choose before Low.

[2] If Y.Y. was not in Oxford in 1920 the first chooser never lent me an umbrella.

[3] If Y.Y. or "Critic" has second choice, "Critic" comes before the one who first fell in love.'

Unfortunately, Low, Y.Y. and 'Critic' could not remember any of the relevant facts; but the family solicitor pointed out that, assuming the problem to be properly constructed (i.e., assuming it to contain no statement superfluous to its solution) the relevant data and order could be inferred.

What was the prescribed order of choosing?

Low, Y.Y. and 'Critic' were colleagues of Phillips on the *New Statesman*, but that hardly helps. The crucial idea is that every piece of information is relevant, so you must exclude all solutions in which any part of any statement is superfluous. Newman's puzzle-*setting* brain would later be put to more serious use puzzle-*solving*. During the Second World War, he ran a section of codebreakers – the Newmanry – at Bletchley Park, which led to the construction of Colossus, the world's first programmable electronic computer. Newman was a colleague and close friend of Alan Turing, the father of theoretical computer science. In fact, Turing's landmark paper *On Computable Numbers* was inspired by Newman's lectures at Cambridge. When Newman set up the Royal Society Computing Machine Laboratory in Manchester after the war, he persuaded Turing to join him there.

Hubert Phillips is the earliest source for this next amazing puzzle: the three-way duel, or 'truel', which I have rephrased in homage to a film that ends in one.

(12)

TRIANGULAR GUNFIGHT

Good, Bad and Ugly are about to take part in a three-way gunfight. Each is positioned on one of the three points of a triangle. The rules are that Ugly will shoot first, then Bad, then Good, before returning to Ugly and continuing in the same order until only one person is left. Ugly is the worst shot and can hit a target only one time in three. Bad is better, hitting a target two times in three, while Good is the best, never ever missing.

You can assume that everyone adopts the best strategy and no one is hit by a bullet that wasn't meant for them.

Who should Ugly aim at to have the best chance of survival?

Here are three more logic puzzles of the kind that Hubert Phillips pioneered, although they were not written by him. They read like one-act plays, and they are just tricky enough to be deliciously satisfying to solve.

(13)

APPLES AND ORANGES

In front of you are three boxes, the first labelled 'apples', the second 'oranges' and the third 'apples and oranges'. One box contains apples, one contains oranges, and the other contains apples and oranges. Each label, however, is on the wrong box. Your job is to correctly reassign the labels. You can't see (or smell) what's in any of the boxes. But you are allowed to stick your hand in one of them and remove a single piece of fruit.

Which box do you choose, and once you see that piece of fruit how do you deduce the correct contents of all the boxes?

(14)

SALT, PEPPER AND RELISH

Sid Salt, Phil Pepper and Reese Relish are all having lunch together when the man among them notices that one of them has picked up the salt, another the pepper and the third the relish.

The person with the salt replies: 'What gives our situation some spice is that no one is holding the condiment that matches their surname!'

'Pass the relish!' adds Reese.

If the man doesn't have the relish, what does Phil have?

(15)

ROCK, PAPER, SCISSORS

Adam and Eve play Rock, Paper, Scissors ten times. It's known that:

- Adam uses three rocks, six scissors and one paper.
- Eve uses two rocks, four scissors and four papers.
- There is never a tie.
- The order that Adam and Eve play their hands is not known.

Who wins and by how much?

When Hubert Phillips died in 1964, his obituary in *The Times* said that 'It may be claimed for him that he provided more amusement for a wet day than any other writer of his time.' As well as publishing puzzles, he compiled thousands of crosswords and wrote extensively on bridge, a game in which he captained England. He also wrote light verse, more than 200 detective stories and an academic treatise on the football pools, and was a much-loved wit on BBC Radio's *Round Britain Quiz*. Yet even though he spread himself across so many fields, his contributions to puzzle culture were as deep as they were extensive.

Phillips was the first person to publish a puzzle involving shared knowledge between participants, which, as we will see, makes him the granddaddy of the Cheryl's Birthday problem that went round the world in 2015.

The earliest of these puzzles involved smudges on faces. The simplest version involves only two people.

MUD CLUB

Alberta and Bernadette are mucking about in the garden.
They come inside. The sisters can see each other's faces,
but not their own. Their father, who can see both girls, tells
them that at least one of them has a muddy face.
 He asks them to stand with their backs to the wall.
 'Please step forward if you have a muddy face,' he says.
Nothing happens.
 'Please step forward if you have a muddy face,' he repeats.
What happens and why?

When solving this type of puzzle we need to assume that all protagonists,
even naughty children, behave honestly and have the analytical skills of a
professional logician.

 I'll take you through it. We know that at least one girl has mud on her
face, so there are three possible scenarios: either Alberta is muddy and
Bernadette isn't, or vice versa, or both of them have mud on their faces.

Case 1. Alberta is muddy, Bernadette is clean. (Note that this is information
known to us as outsiders, not to the sisters. All they know is what they can
see and what they can deduce.)

 Let's enter the mind of Alberta. She looks at Bernadette and sees a clean
face. Since she knows that at least one of them has mud on her face, she
deduces that it must be her. Her father then asks the person with a muddy
face to step forward, but Alberta does not do so. We can therefore deduce
that this scenario must be incorrect, since if Alberta was behaving honestly
she would have stepped forward.

Case 2. Bernadette is muddy, Alberta is clean.

The same logical argument, but with the names swapped, eliminates this scenario too.

Case 3. Both girls are muddy.

Again let's be Alberta. She looks at Bernadette and sees a muddy face. She knows that at least one of them has a muddy face. She cannot deduce anything about the muddiness of her face because in both cases – her having a muddy face or a clean one – the statement that 'at least one of the sisters has a muddy face' is true. So when her father asks the person with the muddy face to step forward, she doesn't. The important point is that Alberta refrains from stepping forward because she is ignorant of the state of her own face – not because she knows her face is clean.

Likewise, Bernadette sees a muddy face, and therefore deduces that she cannot know for sure the state of her own face. When her father asks those with muddy faces to step forward she will also refrain.

We can be sure that this scenario is the correct one, because neither girl moves when questioned by their father for the first time. What happens next?

Alberta either has mud on her face or she doesn't. However, she can eliminate the possibility that she has a clean face because if it was clean, Bernadette, on seeing it, would have already deduced that she had a muddy face, and would have stepped forward the first time their father asked. So Alberta deduces that she has mud on her face. For exactly the same reasons, Bernadette deduces that she has mud on her face, and when their father asks who has a muddy face for the second time both sisters step forward together.

To summarise, this is what happens: both sisters see mud on each other, and therefore cannot deduce information about their own faces. But when they realise that the other sister cannot deduce the state of her own face, they gain new information that allows them to deduce that they both have muddy faces. Neat!

Hubert Phillips published the earliest 'smudge on face' puzzle in 1932, although the logic of face-smudging is more ancient. In the French parlour game *I Pinch You Without Laughing*, which dates at least as far back as the sixteenth century, a person whose fingers are covered in soot makes smudges on the faces of other members of the group. The aim is to be the last to laugh. *I Pinch You Without Laughing* is mentioned in *Gargantua and Pantagruel*, French author François Rabelais' comic masterpiece. An early nineteenth-century German translation of the book describes a novel twist to the game in which everyone pinches their right neighbour on the chin. Two players have their fingers blackened by a charred piece of chalk, so two people end up with smudges on their faces. 'These [players] make a fool of themselves,' notes the translator, 'since they both think that everybody is laughing about the other one.'

Soon after Phillips published his 'smudge on face' problem, variations were soon featuring in puzzle books and attracting the interest of academics. The Russian-American cosmologist George Gamow, one of the earliest advocates of the Big Bang theory of the origin of the universe, was also the author of wonderful science popularisation books. These included *One Two Three ... Infinity* (published in 1947), which is still one of my favourites, and which is especially charming because he illustrated it himself. In 1956 Gamow was consulting for the airline Convair, where the mathematician Marvin Stern had a full-time job. The men worked on different floors and noticed that whenever they went to each other's offices the lift was almost always coming in the wrong direction. They became friends chatting about the maths behind this apparent conundrum, and as a result decided to collaborate on the book *Puzzle-Math*, which contains the following three-person smudge-on-face problem.

(17)

SOOT'S YOU

Three passengers are sitting in a train, minding their own business, when smoke from a passing locomotive suddenly blows through the window, covering all their faces in soot. One of the passengers, Miss Atkinson, looks up from the book she is reading and chuckles. She notices the other two passengers chuckling too. Miss Atkinson, just like her carriage companions, assumes that her face is clean, and that the other two passengers are laughing at the sight of each other's mucky faces. Then she twigs. She takes out a handkerchief and wipes her face.

We can assume that all three passengers are behaving logically, but that Miss Atkinson is quicker on the uptake. How did she work out that her face was also smeared with soot?

Puzzle-Math is not as well remembered as Gamow's other books, but even so it includes one of the most magnificent logic puzzles ever devised. (Gamow credits the puzzle to the great Soviet astrophysicist Victor Ambartsumian.) I have paraphrased and modified it a little, basically by inverting the sexism. It's a hard puzzle, but if you have followed the logic of the previous two problems you will be well equipped to solve it. Even if you don't work it out, you'll be able to follow – and marvel at – the solution.

(18)

FORTY UNFAITHFUL HUSBANDS

In a provincial town 40 husbands are cheating on their wives. Each woman is aware that every man apart from her own husband is having an affair. In other words, each wife assumes her husband is faithful while knowing that all 39 other husbands aren't. On hearing about the moral degeneracy in the town the monarch in the capital issues a decree to punish the husbands for their wickedness. It stipulates that the day after a woman discovers her husband has been unfaithful, she must kill him at noon in the town square.

He then says: 'I know that at least one husband has been unfaithful and I urge you to do something about it.'

What happens next?

The puzzle seems impossible at first, since the wives *already* know of 39 cheating husbands. What difference does it make that the monarch reveals that 'at least one' is a cheat? Yet make a difference it most certainly does!

In a similar vein, this next puzzle involves three people who make deductions based on private and shared knowledge.

(19)

BOX OF HATS

Algernon, Balthazar and Caractacus have a box that contains three red hats and two green hats. They each close their eyes, take a hat from the box and put it on. They close the box and open their eyes, so that each of them can see the colour of the hat worn by the other two. They do not know the colour of their own hat, nor which hats are left in the box.

Algernon says: 'I don't know the colour of my hat.'
Balthazar says: 'I don't know the colour of my hat.'
Caractacus, seeing that the other two both have red hats, says: 'I know the colour of my hat!'
What colour is it?

The Box of Hats puzzle dates from 1940 at the latest, although then it was phrased differently and described disks on the foreheads of Chinese mandarins. More importantly, no mandarin declares his ignorance out loud. One has to deduce what they don't know from their silences.

The comedy dialogue in which each protagonist declares that they don't know something until they do is a winning enhancement that was introduced in the 1960s. The repartee makes it much clearer who knows what and adds to the sense of pantomime.

The following puzzle appears in British mathematician J. E. Littlewood's *A Mathematician's Miscellany* from 1953. Littlewood was one of Britain's three greatest mathematicians in the first half of the twentieth century, along with G. H. Hardy and, so the joke went, 'Hardy-Littlewood', in reference to the incredibly rich and long-standing collaboration between both men. During the First World War Littlewood worked for the army improving the formulae that calculated the direction, time and range of missile trajectories. So valuable was his military work that he was given special allowances, such as permission to carry an umbrella while in uniform.

Back to the puzzle, which is rebooted from Littlewood's original with the now *de rigueur* back-and-forth slapstick. It's challenging because you have to hold in your head the rebounding possibilities as the common knowledge accumulates. The pleasure comes from eliminating what cannot be the case, step by step, to reveal the answer. Logic puzzles compel a clarity of thought that is simultaneously thrilling and pain-inflicting – which is always part of the fun.

(20)

CONSECUTIVE NUMBERS

Zebedee has secretly written down two numbers on a piece of paper. He tells both Xanthe and Yvette that these numbers are whole numbers – i.e., they are taken from the numbers starting 1, 2, 3, 4, 5, ... He also tells them that the two numbers are consecutive – i.e., that they are a number and the next one along, so they are of the form 1 and 2, or 2 and 3, or 3 and 4, and so on. Zebedee then whispers one of the numbers to Xanthe, and the other number to Yvette.

The following conversation ensues:

Xanthe:	'I don't know your number.'
Yvette:	'I don't know your number.'
Xanthe:	'Now I know your number!'
Yvette:	'Now I know your number!'

Can you deduce at least one of Zebedee's numbers?

Rather than whispering a number to Xanthe, Zebedee could have smudged it on Yvette's face, or written it on Yvette's hat. And rather than whispering a number to Yvette, he could have smudged it on Xanthe's face, or written it on Xanthe's hat. What's important to these puzzles is that Xanthe knows something that Yvette doesn't, and vice versa.

This structure underlies the next problem, which I posted on my *Guardian* blog in 2015 after I spotted it on a Singaporean website. It caught my attention because it was described as being for primary schoolers, reinforcing the stereotype of dauntingly brilliant standards in Asian maths education. If Singaporean primary school kids were expected to solve a problem like this one, no wonder they were the best young mathematicians in the world.

(21)

CHERYL'S BIRTHDAY

Albert and Bernard just became friends with Cheryl, and they want to know when her birthday is. Cheryl gives them a list of 10 possible dates.

May 15	May 16	May 19
June 17	June 18	
July 14	July 16	
August 14	August 15	August 17

Cheryl then tells Albert the month and Bernard the day of her birthday.

The following conversation ensues.

Albert: 'I don't know when Cheryl's birthday is,
              but I know that Bernard does not know too.'
Bernard: 'At first I didn't know when Cheryl's birthday is,
              but I know now.'
Albert: 'Then I also know when Cheryl's birthday is.'

So when is Cheryl's birthday?

Within a few hours my post on Cheryl's Birthday was the most viewed story on the *Guardian*'s website. My cheeky clickbait headline – 'Are you smarter than a Singaporean ten-year-old?' – probably helped. Soon after, however, it transpired that the question was taken from a regional maths competition aimed at the top 40 per cent of fifteen-year-olds, and it was the penultimate question in a paper of twenty-five questions of increasing difficulty. So only the very top students would have been expected to get it right. I changed the title to accurately reflect the level of the problem, but interest did not wane. Quite the opposite: the Cheryl's Birthday problem was spreading across the web like a global pandemic. In the following days the logic puzzle was the number one story on many news sites, including BBC and the *New York Times*. The puzzle got more than five million hits that week on the *Guardian* site alone and, when the newspaper came to tally the most viewed stories of the year, the post where I set the problem was in ninth place, and the post with the solution was in sixth. I doubt a maths problem has ever spread so quickly to so many people around the world.

I got in touch with Joseph Yeo Boon Wooi, the Singaporean maths educator who wrote the problem. He only discovered that the puzzle had gone viral when he was surfing Facebook and saw a photograph of the exam paper. 'Hey, this looks familiar,' he exclaimed. 'Wait a minute, I'm the one who set this!' Dr Yeo, of Singapore's National Institute of Education, is the lead author on the maths textbooks used by more than half of secondary school students in Singapore. He told me that the idea for the Cheryl puzzle came from someone else. He had read a similar version of the puzzle online and decided to adapt it, choosing new names for the characters, tightening the dialogue and changing the dates, mischievously making his own birthday the solution. Both he and I have failed to find the original author. We have only been able to trace it to a 2006 post on the Ask Dr Math pages run by Drexel University. The post was submitted by someone called 'Eddie', who was asking for the solution.

One of the lessons of Cheryl's Birthday is that writing a great puzzle is usually a collective enterprise. Like jokes and fables, puzzles change and evolve. Each time a question is rephrased it gains something new, and the best ones can endure for decades, centuries and even millennia.

Joseph Yeo, however, did devise this sequel.

DENISE'S BIRTHDAY

Albert, Bernard and Cheryl become friends with Denise, and they want to know her birthday. Denise gives them a list of 20 possible dates.

17 February 2001	16 March 2002	13 January 2003	19 January 2004
13 March 2001	15 April 2002	16 February 2003	18 February 2004
13 April 2001	14 May 2002	14 March 2003	19 May 2004
15 May 2001	12 June 2002	11 April 2003	14 July 2004
17 June 2001	16 August 2002	16 July 2003	18 August 2004

Denise then tells Albert the month of her birthday, Bernard the day and Cheryl the year.

The following conversation ensues:

Albert: 'I don't know when Denise's birthday is,
          but I know that Bernard does not know.'
Bernard: 'I still don't know when Denise's birthday is,
          but I know that Cheryl still does not know.'
Cheryl: 'I still don't know when Denise's birthday is,
          but I know that Albert still does not know.'
Albert: 'Now I know when Denise's birthday is.'
Bernard: 'Now I know too.'
Cheryl: 'Me too.'

So, when is Denise's birthday?

Another significant ancestor in Cheryl's family history is Dutch mathematician Hans Freudenthal's 'impossible puzzle', from 1969, the first to include the 'I-don't-know-now-I-know' conversation used in the previous problems. Almost true to its name, it is unlikely to be solved with just pen and paper, so I haven't included it here. (But do go online if you're feeling brave.) The impossible puzzle also belongs to another puzzle tradition, which dates back to at least the first half of the last century. In these puzzles we have to deduce a set of numbers from knowing what they add up to (their sum) and what they multiply to (their product). Usually, these problems are phrased in terms of ages. And quite often in terms of men of the cloth.

THE AGES OF THE CHILDREN

The vicar asked the verger, 'How old are your three children?'

The verger replied, 'If you add their ages you get the number on my door. If you multiply their ages together you get 36.'

The vicar went away for a while but then came back and said he could not solve the problem.

The verger told the vicar: 'Your son is older than any of my children', and added that the vicar would now be able to solve the problem.

Find the ages of the children.

Which leads us to the penultimate puzzle of the chapter, devised by the British mathematician John Horton Conway, who is emeritus professor at

Princeton University. The last time I met Conway, at an interdisciplinary maths, puzzle and magic conference, he told an audience of 300 that people like him needed some kind of empowering salute, and he suggested a gesture in which one points at oneself while whimpering 'nerd' as feebly as possible. He then led the room in a mass nerd salute. Conway's playfulness has influenced his entire academic career: he has invented many games and puzzles, most famously the Game of Life. This mathematical simulation of how things evolve which is used by scientists such as Stephen Hawking as a model of how simple rules can produce complex behaviour.

His problem below is a masterpiece. It simultaneously mocks the 'common knowledge' genre while at the same time being a brilliant example of it. Like all the best logic puzzles since Alcuin it presents an amusing story and appears at first glance to give you far too little information to solve it.

(24)

WIZARDS ON A BUS

Last night I sat behind two wizards on a bus, and overheard the following:

A: 'I have a number of children, whose ages are positive whole numbers, the sum of which is the number of this bus, while the product is my own age.'

B: 'How interesting! Perhaps if you told me your age and the number of your children, I could work out their individual ages?'

A: 'No.'

B: 'Aha! AT LAST I know how old you are!'

What was the number of the bus?

When the wizard says no, he is being neither grumpy nor dismissive. He is saying that if he provided his age and number of children then B does not have enough information to deduce the individual ages.

To simplify the solution for you, I can confirm that the wizard has more than one child. There is only one possible number for the bus.

All aboard!

To finish, a visual logic puzzle to get you in the swing for geometry, up next.

Would it help if I mentioned that most people get this puzzle wrong?

(25)

VOWEL PLAY

The following four cards each have a letter on one side and a number on the other.

To verify the following statement is true:

All cards with a vowel on one side have an odd number on the other side.

Which cards do you need to turn over?

Ten tasty teasers

Are you a wizard at wordplay?

1) Add a single letter to either the beginning or the end of the following sequence of letters to make an English word. You cannot change the order of the given letters.

   LYLY

2) The ten letter keys on the top line of a typewriter are:

   QWERTYUIOP

   Can you find a ten-letter word that uses only these keys?

3) Complete the word shown below, using only three new letters. The given letters must appear in that order in the final word, with no letters in between.

   ONIG

4) Jasper Jason works for local radio. This is his business card.

   Can you spot the pattern?

5) Complete the word below. The given letters must appear in that order in the final word, with no letters in between. This time I'm not telling you how many letters you need.

RAOR

6) Puzzle scholar David Singmaster noticed the following pattern while invigilating an exam. He did not accidentally get the T-key jammed on his laptop.

SENTTTTTTTTTTTTTTTTTTTTT

What's the next letter?

7) Complete the word below. The given letters must appear in that order in the final word, with no letters in between.

HQ

8) What do the following words have in common?

Assess
Banana
Dresser
Grammar
Potato
Revive
Uneven
Voodoo

9) Complete the word below. The given letters must appear in that order in the final word, with no letters in between.

TANTAN

10) What letter comes next to complete this sequence?

O U E H R A

A Man Walks Round
an Atom ...

GEOMETRY PROBLEMS

The first book to illuminate the joy of logical deduction was *Elements*, written around 300 BCE by the Greek mathematician Euclid.

Elements is ostensibly about geometry, that is, the behaviour of points, lines, surfaces and solids. Yet its real significance for the history of human thought was the method Euclid introduced to investigate these concepts. The book begins with a set of definitions, and five basic rules he states we can accept as true. He deduces everything else in the book from these original premises, at each stage showing rigorously how each step follows from the one before. The power of this method is that it constructs a huge edifice of knowledge for which the truth of a few preliminary statements guarantees the truth of everything else. The template of *Elements* is the template of all subsequent mathematics.

In practical terms, all Euclid started with was a ruler, for drawing lines, and a compass, for drawing circles. That's it. Every theorem in the book (and there are hundreds of them) is proved using those two instruments alone.

Here, for example, is how you cut a given line in half:

Step 1. Place the compass point at one end of the line, the pencil at the other, and draw a circle.

Step 2. Repeat with the compass point at the other end.

Step 3. Use the ruler to draw a line between the intersecting circles. The line cuts the given line in half.

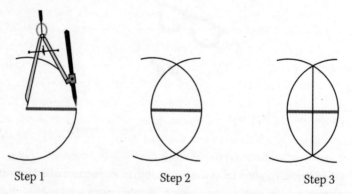

Step 1   Step 2   Step 3

Each theorem in *Elements,* in fact, is presented like a problem, and each proof is presented like its solution. It's a puzzle book in all but name. What I really like about the next problem is that it teases Euclid, the master of conceptual frugality, for having too much equipment in his pencil case.

THE LONE RULER

You have a pencil and a ruler but no compass. The ruler
has two marks on it, as shown below. Can you draw a line
whose length is exactly half the distance between those
two marks? In other words, if the distance between the
marks is 2 units, can you draw a line 1 unit long?

You can measure using only the ruler – not the pencil or
the paper.

The problems I've chosen for this chapter are all geometric in the sense that
they explore, and have fun with, the properties of lines, shapes and physical
objects. The next one originates from an eighteenth-century edition of
Elements that contains notes by William Whiston, Isaac Newton's successor
as Lucasian Professor of Mathematics at Cambridge University. Whiston
noted a mathematical oddity that became a particularly celebrated puzzle.

He imagined a man walking around the circumference of the Earth, and
remarked how much further the man's head would travel than his feet. Can
you work out the distance? Assume that the Earth is a perfect sphere.

I'll do the calculation for you, but first we require a piece of elementary
mathematical knowledge: the formula for the circumference of a circle,
which is twice pi times the radius, usually abbreviated to $2\pi r$, where pi is
approximately 3.14. I'm hoping that introducing a formula here doesn't put
you off what is a delightful and surprising result. Bear with me while we
work it out.

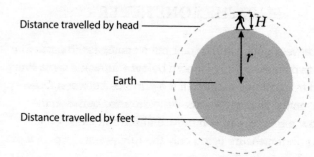

In the diagram above, r is the radius of the Earth and H is the man's height. Using the formula, the circumference of the Earth (the distance travelled by the man's feet) is $2\pi r$ and the circumference of the dotted circle (the distance travelled by his head) is $2\pi(r + H)$, since the radius of the dotted circle is the radius of the Earth plus the man's height. So, the difference between the two circumferences, which is the distance the head travels further, is:

$2\pi(r + H) - 2\pi r = 2\pi r + 2\pi H - 2\pi r = 2\pi H.$

The $2\pi r$ terms cancel out (hold that thought!) and the answer is $2\pi H$, which is $2 \times 3.14 \times$ the man's height.

So, if the man's height is, say, 1.8m, the distance his head travels more than his feet is about 11m.

Now we can see why Whiston thought the answer interesting enough to point out. The distance is tiny! The circumference of the Earth is about 40,000km. It is startling to think that after thousands of kilometres of strolling around the Earth the man's head only travels about 11m more than his feet, or 0.00003 per cent of the total journey.

Whiston's globetrotter is the origin of the following classic puzzle.

(27)

ROPE AROUND THE EARTH

A rope lies tightly around the circumference of the Earth. The rope is then extended in length by 1m and raised up at every point from the ground until it is again a taut circle, and every point on the rope is the same height above the ground.

How high is the rope now? What size of animal could crawl under it?

The illustration below shows how this problem is essentially the same as the previous one. Both involve the comparison of two concentric circles, the smaller of which is the circumference of the Earth. In the rope's case the bigger circle has a circumference that is 1m longer than the smaller circle.

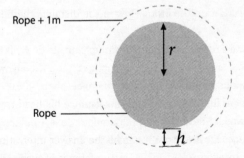

When the problem is phrased in terms of the rope, the counter-intuitive answer is even more powerful. By extending the rope by 1m, we will be able to raise it off the ground by $\frac{1}{2\pi}$ m, which is about 16cm. (My workings: let

c be the circumference of the Earth, so the length of the extended rope is $c + 1$. Using the formula for circumference we have two equations: $2\pi r = c$ and $2\pi(r + h) = c + 1$. These combine to get $2\pi h = 1$, or $h = \frac{1}{2\pi}$.)

Think about this result for a second. We have a rope that is 40,000km long, and we extend it so it is 40,000.001km long. Yet this apparently insignificant increase creates enough slack for the rope to stand 16cm off the ground all the way around the world. What animal will be able to get through? Easily a cat or a small dog.

Now let's return to that thought, from our discussion of the man walking around the Earth. When we worked out the extra distance the head travels, the two $2\pi r$ terms cancelled out, leaving the answer as 2π times the man's height. What is significant here is that the Earth's radius, r, appears nowhere in the answer, meaning that the extra distance travelled by the head is determined *only* by the man's height and not by the size of the Earth. In other words, the size of the Earth makes no difference to the answer. Whiston's rambler could be walking around any sphere at all, and his head would always cover an extra 11m.

(1) A man walks round an atom. How much further does his head travel than his feet?

(2) A man walks round a football. How much further does his head travel than his feet?

(3) A man walks round Jupiter (circumference: about 400,000km). How much further does his head travel than his feet?

(4) A man walks round the Sun (circumference: 4.4 million km). How much further does his head travel than his feet?

The answer in all cases is only 11m (ignoring, of course, the physical challenges involved). Likewise, if a rope was girdling an atom, a football, Jupiter or the Sun, an extra metre in length would create enough slack to raise the rope by 16cm at all points. Amazing.

William Whiston lasted only eight years as Lucasian Professor before he was expelled from Cambridge for heresy. (His offence was to reject the idea of the Holy Trinity, arguing instead that Jesus was subordinate to God.) Whiston never returned to academia, instead giving maths and science lectures in London's coffee houses, where he had the tendency to digress into religious rants.

Whiston's greatest contribution to science was the crucial role he played in persuading the British government to establish the Board of Longitude, which offered prize money to the first person to invent a way to determine a ship's longitude at sea. Ever hopeful to win the cash, he nevertheless failed in all his attempts to solve the problem. It is beautifully fitting, therefore, that his greatest contribution to maths was a puzzle about circumnavigating the Earth.

I prefer Whiston's phrasing of the problem as a man walking around the Earth to the later version of a rope hovering above the ground, since – although both are obviously absurd – the former scenario seems less contrived. If there was a rope girdling the Earth, and you *did* extend it by 1m, before even thinking of attempting to levitate it everywhere, surely you would raise it at a single point to see how far it would go. Especially if the purpose was to let animals pass under it!

New problem:

> *You have a rope that is lying around the circumference of the Earth. You extend it by a metre. Now you raise it upwards at a single point until the slack becomes taut. How high does it go? What animal will fit under it now?*

Don't struggle with the calculation, because the answer is only accessible to those with a certain level of maths. I have included the question here since the answer is interesting. Have a guess, and peek at the answer in the back.

In fact, do the next question first, and *then* peek at the answer.

Clue: we will need Pythagoras's theorem, which states that for all right-angled triangles the square of the hypotenuse is equal to the sum of the squares on the other two sides. (The hypotenuse is the side opposite the right angle.) But you knew that anyway, right?

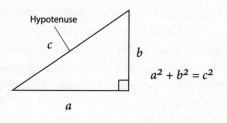

$$a^2 + b^2 = c^2$$

(28)

BUNTING FOR THE STREET PARTY

There will be a party on your street, which is 100m from end to end. You have a 101m line of bunting for the decorations. You attach one end of the bunting to the bottom of a lamp post at one end of the street, and the other end to the bottom of a lamp post 100m away at the other end of the street. You hang the middle of the bunting from the top of a pole halfway down the street.

Assuming there is no slack and no stretching of the bunting, how high will the pole be?

The next three puzzles concern the behaviour of rolling circles. They might spin your head if you have never thought about these ideas before, but I guarantee you will appreciate the 'wow' however you discover the answers. The puzzles will become more accessible after that, when we visit Japan.

Elements established Euclid as the laureate of logic, the high priest of cold and rigorous deductive thought. This reputation is now shared, and possibly eclipsed, by Sherlock Holmes.

The fictional sleuth aspired to Euclidean rigour – 'How often have I said to you that when you have eliminated the impossible, whatever remains, *however improbable*, must be the truth?' – but he was not as good at maths.

In one early Sherlock Holmes caper, *The Adventure of the Priory School*, Holmes looks at the wheel tracks made by a bicycle and deduces which way it was heading. He explains his reasoning to Watson: 'The more deeply sunk impression is, of course, the hind wheel, upon which the weight rests. You perceive several places where it has passed across and obliterated the more shallow mark of the front one. It was undoubtedly heading away from the school.'

I'm not sure I follow. Surely the back wheel would obliterate the front wheel's mark in whichever direction the cyclist was going?

Holmes creator Sir Arthur Conan Doyle missed a trick. It *is* possible to deduce the direction of travel of a bicycle from its tyre tracks.

(29)

ON YER BIKE, SHERLOCK!

Was the cyclist who left the tracks below going from left to right, or from right to left?

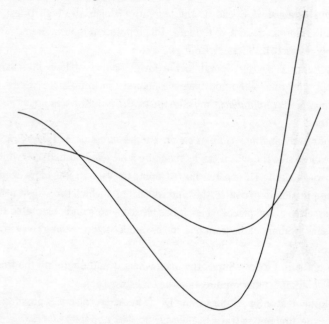

Holmes was correct that you first need to work out which wheel made which track, but you can do this without knowing how deep each impression is.

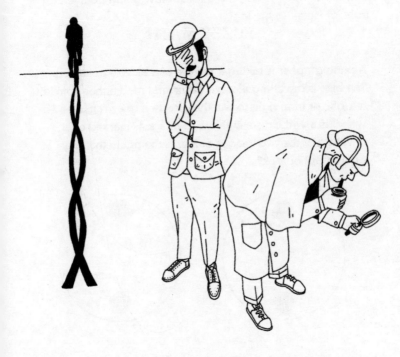

Here's another bicycle puzzle. You may instinctively know the answer. One image feels right, and the other doesn't. But can you work out why?

(30)

FUZZY MATH

A photographer is taking a picture of a bicycle in motion. The bike is travelling along a horizontal road either from left to right, or from right to left; the direction doesn't matter. The wheel is a white disc, with two pentagons marked on it.

Which of the two images below is the photo that the photographer took?

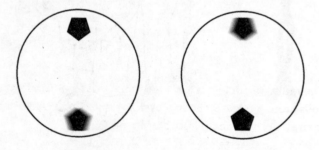

The lesson of the above puzzle is that the behaviour of a rolling circle is subtler than at first it appears.

The following question is taken from a SAT general aptitude test that 300,000 Americans took in 1982. Only three students got the correct answer. Will you?

ROUND IN CIRCLES

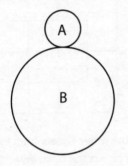

The radius of circle A is $\frac{1}{3}$ of the radius of circle B. Circle A rolls around circle B one trip back to its starting point. How many times will circle A revolve in total?

(a) $\frac{3}{2}$

(b) 3

(c) 6

(d) $\frac{9}{2}$

(e) 9

Now for something to scrunch your mind in a different way.

(32)

EIGHT NEAT SHEETS

Eight sheets of identically sized square paper are placed on a table. Their edges form the following pattern, with only one sheet, marked 1, completely visible.

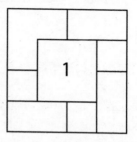

Can you number the other sheets from top to bottom where 2 is the second layer, 3 is the third layer, and so on?

I first read about the neat sheets in Kobon Fujimura's brilliant *The Tokyo Puzzles*. Between the 1930s and the 1970s, Fujimura was Japan's puzzle king. He published many books, some bestsellers, and in the 1950s he even had his own weekly TV puzzle show. Fujimura's popularity presaged the modern boom in Japanese pencil puzzles, epitomised by the international success of Sudoku in the 2000s, which I'll discuss in more detail later in this chapter.

The Japanese have a more playful approach to numbers than we do in the West, or at least this is how it has felt to me the two times I have visited Japan. Schoolchildren recite their times tables with the happy levity of a nursery rhyme. A popular pastime used to be to play games with the numbers on your subway ticket. And the country has turned mental arithmetic into a spectator sport. Learning the abacus is still a popular after-school activity, and there is a tournament circuit for its best practitioners. When I attended the national abacus championship in 2012, the climax was a counting game in which contestants visualising an abacus had to add fifteen numbers flashed at them in under two seconds. It was tense and exciting!

Here's another of Fujimura's puzzles that I really love.

A SQUARE OF TWO HALVES

A big square is divided into sixteen smaller squares. The images below show two ways to cut this big square into two identical pieces.

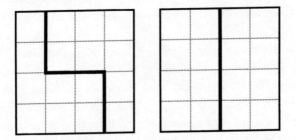

There are four more ways to do this. Can you find them?

For clarification, you can only cut along the inner lines and the two shapes must be identical, meaning that if they were made of card you would be able to make them align perfectly by sliding one on top of the other and adjusting their positions while always keeping them flat. If you need to turn one shape upside down – that is, take the card off the table and replace it with the top side on the bottom – in order to make it align perfectly with the other card, the shapes are not identical.

Finally, here is a Fujimura puzzle with curves. You may need the formula for the area of a circle, which is pi multiplied by the square of its radius, or πr^2.

(34)

THE WING AND THE LENS

Illustrated below is a quarter circle that contains two smaller semicircles. Prove that the wing shape A has the same area as the lens shape B.

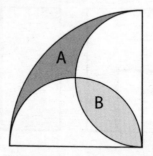

I like this puzzle not only because the image is pretty, but also because it reminds me of the Japanese tradition, between the seventeenth and nineteenth centuries, of hanging decorative wooden tablets containing geometry problems outside shrines and temples. Called *sangaku*, these mathematical images were both religious offerings and public announcements of the latest discoveries. They made mathematics a public event, a source of visual entertainment and wonder. I saw a sangaku tablet at a temple in Kyoto. It included pictures of circles, triangles, spheres and other shapes, beautifully painted in white and red. The composition of geometrical shapes in a sangaku is harmonious and artistic, an aesthetic noticeably absent from the purely didactic images you might find in a Western geometry text book. A sangaku tablet usually has just the final image of a problem, with minimal inscriptions underneath. Many hundreds of sangaku survive, such as the one overleaf, dated 1865, from a temple near Nagoya. The problem is credited to a fifteen-year-old boy, Tanabe Shigetoshi.

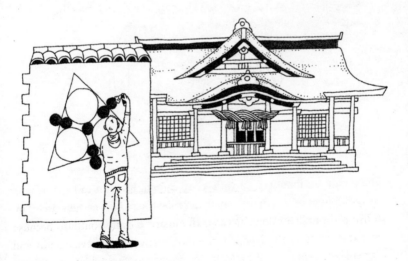

(35)

SANGAKU CIRCLES

The image below shows five sizes of circle. From smallest to largest there are six white circles, seven dark grey ones, three light grey ones, one with a dotted line that sits in the triangle and a larger one with a solid line.

How many radii of the white circle fit along the radius of the circle with the dotted line?

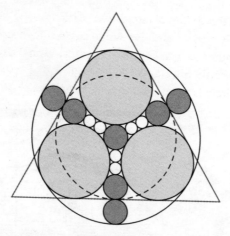

The puzzle bamboozles you with its splendour. It's hard to know where to start. But once you find a way to describe the radii of certain circles in terms of the radii of other circles, you'll discover a really beautiful puzzle.

An even younger Japanese teenager devised the next problem. Thirteen-year-old Sato Naosue's sangaku was hung in 1847 at a temple about 300 miles north of Tokyo. It is trickier than the last one because – as in almost all problems with right-angled triangles – you need to know Pythagoras's theorem. (For a refresher on the theorem go back a few pages to page 58.)

(36)

SANGAKU TRIANGLE

Below are three sizes of circle: two black, three white and one grey. Two black and two white circles are inscribed inside a square, which is itself – together with the other circles – inscribed inside a triangle. Show that the radius of the grey circle is twice the radius of the black circle.

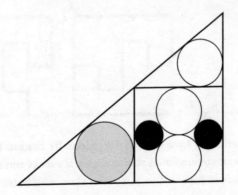

Another Japanese tradition is the use of *tatami mats*. Made from woven straw, and so soft that you walk on them without shoes or slippers, they are usually rectangular in shape and twice as long as they are wide.

(37)

TREADING ON THE TATAMI

Below is an arrangement of tatami mats. Imagine you are walking from A to B along the edges of the mats. If you want to take the longest path, one idea is to start by following the longest possible straight line – along the top, say, as in the middle image; or down the side, as in the right-hand image.

But there is an even longer path than these two. Can you find it?

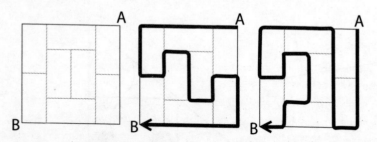

If you ever want to lay tatami mats, you need to be aware that there is an auspicious and an inauspicious way to do so. The lucky way – as demonstrated in the previous question – is to lay them so that whenever three mats meet they make the shape of a 'T'. The unlucky way is to have four mats meeting at the corners, which makes a '+'. Lucky layouts never have four mats meeting at a single point. This superstition makes for some really fun puzzles.

FIFTEEN TATAMI MATS

Tile the room below with fifteen 2 × 1 tatami mats, according to the rule that no four mats can meet at a corner.

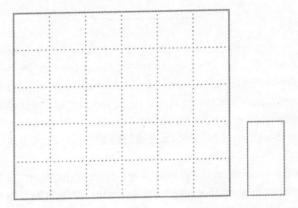

In this question and the next one, use a pencil so you can rub out wrong turns.

Nob Yoshigahara was a Japanese chemical engineer who turned to writing puzzles after he was burned by a chemical explosion. By the time he died in 2004 he was one of the world's most influential puzzle people: a columnist, author, toy designer, collector and international conference organiser. He is remembered by his friends in the global puzzle fraternity as a charismatic, generous and eternally playful figure. His most successful toy, Rush Hour, a logic game in which you must slide plastic cars and lorries around a grid, has sold upwards of ten million copies around the world.

Yoshigahara devised the 'number tree' puzzle that begins this book. He also introduced a new twist to tatami-laying problems. In the pattern below, a straight line (in bold) goes from one side of the room to the other. The next puzzle is about a room where no lines cross the room.

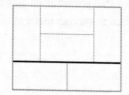

(39)

NOB'S MATS

Tile the room from the previous question with fifteen 2 × 1 tatami mats such that no single line goes straight across the room. Mats are allowed to meet four at a corner.

But rooms are not always rectangles! In the problem opposite, two corner squares are taken up by staircases.

(40)

AROUND THE STAIRCASES

When the room from the previous two problems has opposing corners cut off, it is possible to cover it with 14 tatami mats without gaps or overlaps, as shown below left. (The mats can be placed in any position.) Let's extend the room so it is now 6 × 6, still with two corners cut for the staircases. Prove that you cannot cover the new room with 17 mats without gaps or overlaps.

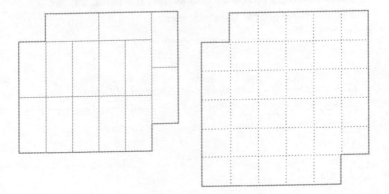

But staircases are not always in corners! In the problem overleaf, the positions of the two staircases are chosen at random.

(41)

RANDOM STAIRCASES

Architects have decided that they do not want to put two staircases in opposing corners of our 6 × 6 room. If the squares in the room are shaded like a chessboard, as shown below, and if one staircase is in any white square and one staircase is in any shaded square, show that it is possible to tile the room with 17 tatami mats leaving neither gaps nor overlaps. Mats cover two adjacent squares and can be placed in any position as long as they are not in the two squares removed for the staircases.

In this puzzle you need to show that it is always possible to tile the room, not just provide an example of when it does.

When I posted the next problem in my *Guardian* column, several architects mocked me for how simple they found it, since the solution is a common design feature of British homes. Their response was the clearest reminder that puzzles that mangle the brains of some solvers can sometimes be almost embarrassingly obvious to others.

(42)

WOODBLOCK PUZZLE

Below are the top and front views of a three-dimensional wooden structure with flat sides. Draw at least one view of the structure from the left-hand side.

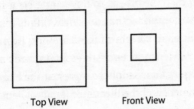

Top View   Front View

All visible edges are marked as lines. Hidden edges must be marked as dotted lines. So, for example, the object below of two square sides with square holes in them, joined at an edge, is not a solution because the side, top and front views of this object would have dotted lines for the hidden edges, as shown below. Of course, it is fine for a side view to have hidden edges. But there can be no hidden edges on the top or front views, since that contradicts the images in the question, which have no dotted lines.

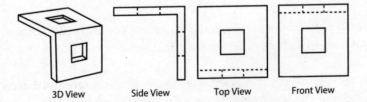

3D View   Side View   Top View   Front View

It is important to note that since the object is made of wood, no part of it can have zero thickness.

The next two puzzles take us inside the home.

The Borromean rings are a fascinating mathematical object: three interlocking rings that have the intriguing property that even though they are all linked, when any single ring is taken away, the remaining two are no longer linked, as illustrated opposite. (If you make the rings from rigid material, the way they overlap forces each ring to face in a slightly different direction than the others, so the diagram is a slight cheat.) I find it pleasingly counter-intuitive that no single ring is joined to any other, but that they are inseparable when taken together. The rings are a popular symbol for the interdependence of three parts, and have been used in Christian iconography, for example, to represent the Holy Trinity.

The rings are named after the Borromeo family of Renaissance Italy, who had an image of three interlocking rings on their coat of arms, although the idea of three objects connected in that way pre-dates them. The *valknut* is a Viking emblem with three interlocking triangles, now most commonly featured in tattoos, pendants and heavy metal T-shirts.

The Borromean rings have three linked elements that fall apart completely when any one of the parts is removed. The same idea is behind the next puzzle.

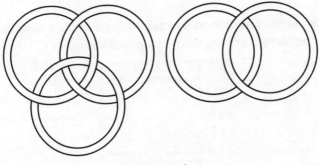

Borromean rings

Valknut

(43)

PICTURE ON THE WALL

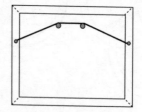

The normal way to hang a picture with two nails is to let the string hang over both nails, as shown below.

The advantage of two nails is that if one nail comes off the wall the picture stays up, because the string is supported by the other nail.

Can you think of a way to loop the string around the nails such that if any one of the nails fails, the picture will also fall to the floor? (You can increase the length of the string if you need to.)

Rings and home furnishings lead us naturally to the mathematical idea of a 'napkin ring': the shape you get when you drill a cylindrical hole through a sphere, where the centre of the hole passes through the centre of the sphere. The following is an incredible puzzle because it contains so little information.

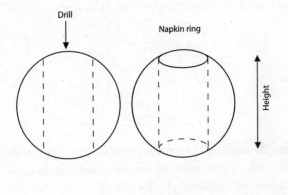

A NOTABLE NAPKIN RING

A napkin ring is 6cm deep. What is its volume?

There is a fair bit of legwork here, but don't let that deter you. I'll start you off. Trust me, this is an amazing puzzle.

The volume of the napkin ring is the volume of the sphere minus the volume of the central bit that got removed, which looks like a cylinder with domes on the top and bottom.

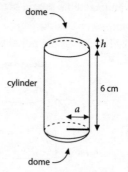

I've added the height of the cylinder, which is 6cm. Let the radius of the sphere be r, the dome's height be h and the radius of the cylinder cross-section, which is also the radius of the base of the dome, be a. Now all that you require are the formulae for volume, which I will generously supply here.

The volume of the sphere: $(\frac{4}{3})\pi r^3$

The volume of the cylinder: $\pi a^2 \times 6\text{cm}$, or $6\pi a^2$

The volume of each dome: $(\frac{\pi h}{6})(3a^2 + h^2)$

We are almost there. The volume of the napkin ring is equal to the volume of the sphere minus the volume of the cylinder minus two times the volume of the dome. Using Pythagoras's theorem we can write a in terms of r, and we can also write h in terms of r. So, we can write the volume of the ring as an expression in which the only variable is r. The expression will be long and crammed full of rs and πs ...

What are you waiting for?!

The historian Herodotus wrote that geometry originated with the Egyptian practice of measuring areas of arable land flooded by the Nile. Calculating the areas of squares and rectangles is still one of the first tasks we learn in geometry: all you do is multiply one side by the neighbouring side.

This very simple procedure is all you need to solve Menseki Meiro, or Area Maze, a wonderful puzzle by the Japanese inventor Naoki Inaba.

Here's a sample question, so you get the hang of it. Your task is to find the missing value. The marked distances are not exact, so you cannot get the answer by measuring.

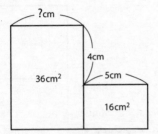

The beauty of this puzzle is that you must solve it geometrically, and with whole numbers. You are not allowed to spoil your workings with equations or – heaven forbid – fractions. To solve this Area Maze, complete the large rectangle as shown below. The area A must be 20cm, since it is 4cm × 5cm. Which means that the area of A plus the rectangle below it is 20cm + 16cm = 36cm. This area is the same as the large rectangle on the left side. Since they share the same height, they must share the same width – so the missing value is 5cm.

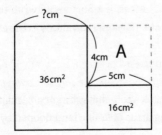

(45)

AREA MAZE

Find the missing value.

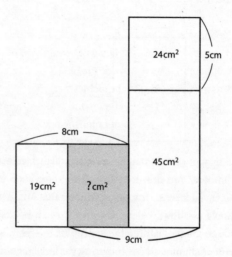

Naoki Inaba is probably the most prolific and brilliant designer of deduction puzzles working today, although his work is barely known beyond his home country. Japan, in fact, sustains arguably the most vibrant puzzle community in the world, through individuals like Inaba and the magazine publisher Nikoli.

You probably haven't heard of Nikoli, but you will have heard of Sudoku, which first appeared in the magazine *Puzzle Communication Nikoli* in the mid-1980s. Sudoku was Nikoli's rebranding of a puzzle called Number Place from the American magazine *Dell Pencil Puzzles and Word Games*. In case you have been living under a rock (or tatami mat) for the last decade, the puzzle is a 9 × 9 grid with some given digits. The solver must fill the blanks so that each digit appears once in each row and column and also once in each of the 3 × 3 subsquares within the main grid. Sudoku didn't attract much attention until 1986, when Nikoli decided to position the given digits in a symmetrical pattern, like letters in crosswords. The tweak worked, and it became a big domestic success. In late 2004, Sudoku appeared in the West for the first time, after an English speaker, Wayne Gould, who had spotted the puzzle while on holiday in Japan, offered Sudoku grids he had made using a computer program to newspapers including *The Times* in London. Within a few months of Sudoku's first appearance in *The Times*, the puzzle was a permanent daily feature in many newspapers around the globe.

It is ironic that Nikoli is best known for a puzzle it didn't create, because since the launch in 1980 of its quarterly magazine the company has published about 600 new types of puzzle. Its speciality is grid puzzles, like Sudoku, in which a grid, usually square, needs to be filled in. Part of the puzzles' charm is the attention to detail – the grids are presented beautifully with the elements arranged symmetrically, or, if not, with care given to how they look. The rules are always very simple, and the pleasure of using a pencil to gradually fill in the spaces is satisfying, addictive and, to people like me, as therapeutic as colouring in. I've chosen four examples to seduce you.

Nikoli's magazine has a circulation of about 50,000. Its readers are not only solvers but also the creators of many of the puzzles, sending in several hundred suggestions a year. The next in our collection, Shikaku, was an idea sent to Nikoli by 21-year-old university student Yoshinao Anpuku. Later, he joined the company, and is now its executive editorial officer.

(46)

SHIKAKU

In Shikaku the aim is to divide a grid into rectangular and square 'boxes'. The numbers in the grid determine the area of the box containing that number (as measured by the number of cells).

I'll work through an example with you. In the illustration below, A is the opening grid and C the completed solution, in which all the rectangular and square boxes are marked. A good way to start is to look for the largest number in the opening grid, since often the shape and position of its box will be restricted. The largest number in this puzzle is 9, and the only possible boxes for a 9 are a 9 × 1 rectangle or a 3 × 3 square. There are no horizontal or vertical lines of nine empty cells, so the box must be the square, and it fits in only one position, illustrated in B. Likewise the only rectangular box of 8 cells that contains the 8, and the only rectangular box of 6 that contains the 6, must be in the positions marked. Once some boxes are drawn in, you can deduce the positions of other boxes.

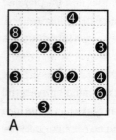
A

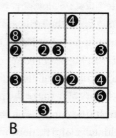
B

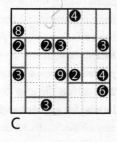
C

Now it's your turn.

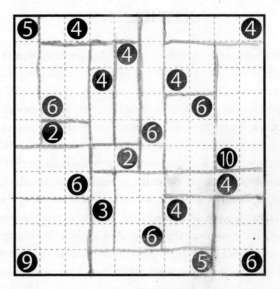

Nikoli was founded by Maki Kaji, a horse-racing obsessive who named his company after the Irish-trained colt that was the losing favourite in the 1980 Epsom Derby. I first met Kaji at Nikoli's offices in Tokyo in 2008. He told me about two of his hobbies: collecting rubber bands and taking photos of licence plates which display lines in the times tables, such as 23 06 (2 × 3 = 6) and 77 49 (7 × 7= 49). Catching up with him again in 2016, he said his rubber band collection continues to expand, with notable new items from Thailand and Hungary. And he now has pictures of about 85 per cent of the lines from the one to nine times tables. 'I'm nearing completion. But I have a rule: that I won't intentionally look for them. I take photos of the ones I come across incidentally.'

(47)

SLITHERLINK

In Slitherlink, the aim is to connect the dots horizontally and/or vertically and to make a single connected loop. The numbers in the grid tell you how many lines surround that number. So, there will be one line around 1, two lines around 2, and so on. You do not know how many lines surround squares with no digits in them. The final loop must never cross itself, nor branch off.

In image A below, the obvious places to start are the 0s, since no lines go around them. Mark little crosses around the 0s to show that no lines can go there, as in image B. One of the crosses is by a 3, leaving only three possible spaces for lines around that number. Fill those lines in. In Image C we can continue the loop upwards, and there is only one way it can go around the 2. Note that in image B I also marked a, b, c and d in the spaces where potential lines could come from the dot between two 3s. The loop must pass through this point, since when three lines surround a cell all four dots are used. The loop must pass through either a or b, and either c or d, since if it passes through both a and b or both c and d then the loop will have a branch, which is forbidden. In all cases it must pass through the other two sides of both 3s, so we can mark them in on image C. The full loop is illustrated in image D.

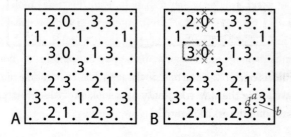

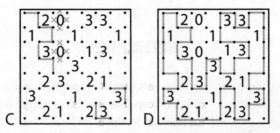

Here's one for you. Remember there is a single loop, with no crosses or branching. There is only one solution and it can be deduced by logic alone.

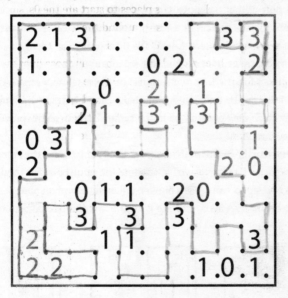

Slitherlink is one of Kaji's favourite Nikoli puzzles and it's one of my favourites too. I love the way you slowly construct the loop, which gradually snakes around the grid. Filling it in is immensely gratifying.

(48)

HERUGOLF

Nikoli is constantly introducing new puzzles. Herugolf is a recent offering that was inspired by golf, and how your shots tend to get shorter and shorter as you approach a hole.

In Herugolf you must putt each ball, denoted by a circle, into its correct hole, denoted by an H. The number in the circle is the distance in cells that the first putt of that ball must travel. If the first putt does not reach an H square exactly, the second putt must travel one cell less. If the second putt does not reach an H square exactly, the third putt must travel one cell less, and so on. So, the path of a 3-ball will either travel three squares exactly, or three squares followed by two squares, or three squares followed by two squares followed by one square. The balls can only travel horizontally or vertically. Each new putt can be in the same direction as the previous one, or change direction.

The paths of the balls cannot cross, and balls must reach the holes exactly at the end of each putt. No two balls go into the same hole. The shaded areas are bunkers. A putt can cross a bunker, but cannot land on one.

In the example, A is the starting grid. The first thing to do is to see if any

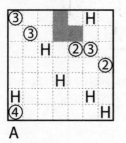

A

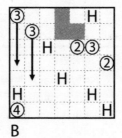

B

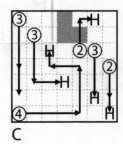

C

balls have their first putt determined. The 3-ball in the top left corner must travel three squares on its first putt. But it cannot go horizontally since it will land on a bunker, so it has to travel vertically down, as in image B. Likewise the 3-ball diagonally adjacent to it will also land in a bunker if putted horizontally, so it must travel down. Each of these balls must continue with a putt that goes only two squares, so the ball from the top left corner is forced to continue down, since the paths cannot cross, and the ball reaches the hole. The other ball must move horizontally, because if it goes down there is no hole it can reach with its following putt, which will be a single square. So it also reaches its hole. Image C shows the completed grid.

Now tee off!

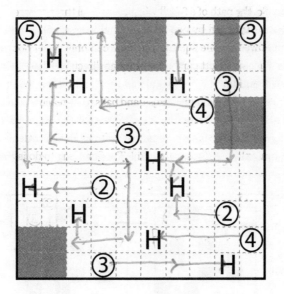

(49)

AKARI

My final Nikoli puzzle is also inspired by the real world – in this case bulbs lighting up a room. The aim of Akari is to illuminate the entire grid by placing light bulbs, which are drawn as circles. A black cell with a number indicates how many bulbs need to be positioned in the neighbouring cells either directly above, below, to the left or to the right of that number. Each bulb lights up all unblocked squares in its row and column. Squares that are not adjacent to numbers may or may not have bulbs in them. In the final grid all white squares must be illuminated, and no two bulbs can be in each other's light path.

In the example, the empty grid is A. Since each number indicates how many bulbs are next to it vertically and horizontally, we know that there are bulbs in all horizontal and vertical positions next to the 4, and also that there are no bulbs in the horizontal and vertical positions next to both 0s, which I have marked with dots in image B. Since two of the bulbs adjacent to the 4 are also adjacent to the 2, we know that the other sides of the 2 cannot have any bulbs, so I have added an extra dot in the square above the 2. In image C, the arrows show the rows and columns illuminated by the four bulbs placed in the previous image. The square above the 3 cannot

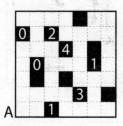

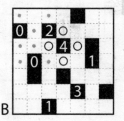

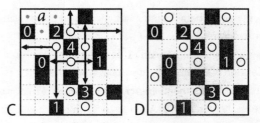

C   D

have a bulb because it is in the path of another bulb, so the other three sides of the 3 must each have one. We can also deduce that there has to be a bulb in *a*, since all other positions that would illuminate that square are forbidden from having bulbs, either because we marked them as having no bulb or because they are in the light path of another bulb. The completed puzzle is shown in D.

Now shed some light on these proceedings.

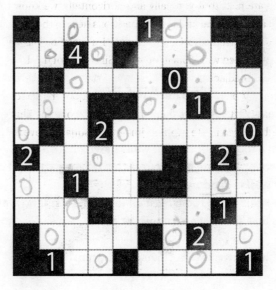

The placement of light bulbs in interestingly shaped rooms leads us to our final geometrical puzzle.

The image below is the horizontal cross-section of a room. The bulb shows the position of the single light source. When the light is switched on, the wall marked in bold remains completely in shadow. (We're assuming no reflections from the other walls).

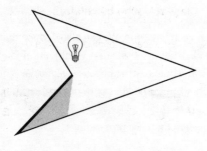

50

THE DARK ROOM

Design a room with straight walls in which there is a position for a single light source that leaves part or all of every wall in shadow.

The walls of the room must all join. Free-standing walls or edges that stick out are prohibited.

Would the last person to solve this one please turn off the lights.

Ten tasty teasers

Are you smarter than a 12-year-old?

Class rules: no calculators allowed.

1) Four of these jigsaw pieces fit together to form a rectangle. Which one is not used?

2) If the following fractions are arranged in increasing order of size, which one is in the middle?

A $\frac{1}{2}$ B $\frac{2}{3}$ C $\frac{3}{5}$ D $\frac{4}{7}$ E $\frac{5}{9}$

A<C   A<D
A<B   A<E

3)

This sentence contains the letter e _____ times   8

seven   eight   <u>nine</u>   ten   eleven

How many of the five words shown above can be placed in the gap to make the sentence in the box true?

A 0   B 1   C 2   D 3   E 4

4) The diagram below shows a Lusona, a sand picture drawn by the Tchokwe people from the West Central Bantu area of Africa. To draw a Lusona the artist uses a stick to draw a single line in the sand, starting and ending in the same place without lifting the stick in between. At which point could this Lusona have started? (At an intersection of lines, the broken line is the one that is drawn first and the unbroken line is the one drawn over it.)

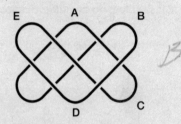

5) Which of the following can be divided by all the whole numbers from 1 to 10 inclusive?

A 23 × 34   **B** 34 × 45   **C** 45 × 56   **D** 56 × 67   **E** 67 × 78

6)

Knave of Hearts: 'I stole the tarts.'

Knave of Clubs: 'The Knave of Hearts is lying.'

Knave of Diamonds: 'The Knave of Clubs is lying.'

Knave of Spades: 'The Knave of Diamonds is lying.'

How many of the four knaves are telling the truth?

A 1   **B** 2   **C** 3   **D** 4   **E** more information needed

7) The faces of a cube are painted so that any two faces which have an edge in common are painted different colours. What is the smallest number of colours required?

A 2 **B** 3 **C** 4 **D** 5 **E** 6

8) Granny swears that she is getting younger. She has calculated that she is four times as old as I am now, but remembers that five years ago she was five times as old as I was then. What is the sum of our ages now?

A 95 **B** 100 **C** 105 **D** 110 **E** 115

9) In the following expression, replace each ☐ with either + or − so that the result of the calculation is 100.

123 ☐ 45 ☐ 67 ☐ 89

The number of + signs is p and the number of − signs is m. What is the value of $p − m$?

A −3 **B** −1 **C** 0 **D** 1 **E** 3

10) The tiling pattern shown below uses two sizes of square, one with sides of length 1 and one with sides of length 4. A very large number of these squares is used to tile an enormous floor in the pattern shown below. Which of the following is closest to the ratio of the number of grey tiles on the floor to the number of white tiles?

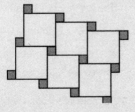

A 1:1 **B** 4:3 **C** 3:2 **D** 2:1 **E** 4:1

Mathematics Most Fowl

PRACTICAL PROBLEMS

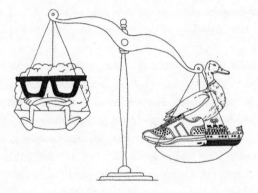

In this chapter I have collected puzzles that take place in the real world. Some involve familiar objects, such as bowls, jugs, fuses, cars and potatoes. Others describe situations from daily life, such as running in a race, flying in a plane and, to start with, the oldest problem in this book, about going to the shops.

(51)

ONE HUNDRED FOWLS

If a cock costs 5 pieces, a hen 4 pieces and a chick $\frac{1}{4}$ of a piece, how many cocks, hens and chicks can be bought with 100 pieces to make 100 of them in all?

The Chinese mathematician Zhen Luan posed this question in the middle of the sixth century, although the style of question – 100 of three types of animal for 100 units of currency – first appeared a century before that, also in China.

It's a great puzzle: brilliantly concise, with an answer that's not immediately evident. Try out a few numbers in your head and you will be bamboozled. Beloved of the Chinese, 'hundred animals' problems spread to India, the Middle East and Europe. Alcuin's *Propositiones*, the eighth-century puzzle trove I mentioned in the chapter on logic problems, includes three versions: boars, sows and piglets at ten, five and half a denarius; horses, cows and sheep at three, one and a twenty-fourth of a solidus; and camels, asses and sheep at five, one and a twentieth of a solidus. The last one he calls the 'puzzle of the oriental merchant', possibly paying tribute to its origins in the East.

Modern readers approaching these problems will instantly turn them into equations. If you are buying x cocks, y hens and z chicks, Zhen Luan's question can be rephrased as:

(1) $x + y + z = 100$ (since the total is 100 animals)

(2) $5x + 4y + z/4 = 100$ (since the money adds up to 100 pieces)

Solve them, and you have your answer.

Zhen Luan, Alcuin and their respective contemporaries, however, solved these problems using guesswork, trial and error. They were not able to use algebra, because algebra had not yet been invented.

It's much more straightforward – and arguably more fun – to solve the puzzles using equations. In fact, what I love about 'hundred animals' problems is that they were among the earliest types of problem to demonstrate the tremendous power of algebraic methods. The puzzles played a role in the development and propagation of these new mathematical techniques, not only as a brilliant advert for their effectiveness, but also as interesting problems that led to deeper analysis by medieval and Renaissance mathematicians.

Algebra is the branch of mathematics in which numbers and quantities are represented by symbols – xs, ys and zs – in equations. The word comes from the Arabic *al-jabr*, meaning restoration. The ninth-century Baghdadi scholar Al-Khwarizmi used this word to mean the mathematical operation understood in modern terms as taking something from one side of an equation and 'restoring' it to the other side. Using *al-jabr* and other proce-dures, Al-Khwarizmi developed methods for solving simple equations. Abu Kamil, an Egyptian mathematician born in the ninth century, wrote some of the earliest treatises that expanded on Al-Khwarizmi's ideas. One of them was about problems involving the purchase of 100 birds for 100 units of currency. 'I am acquainted with a type of problem which proves to be engrossing, novel and attractive alike to high and low, to the learned and the ignorant,' he wrote. 'But when others discuss solutions with one another, they exchange inaccuracies and conjectures, as they see no evident principle or system ... So I decided to write a book to make the matter better understood.'

Let's solve the question in hand. We have two equations:

(1) $x + y + z = 100$

(2) $5x + 4y + z/4 = 100$

To solve equations like these – called *simultaneous equations* in school – you usually need as many equations as there are variables. So, for three variables, we'd need three equations.

Here, we only have two. But the question provides us with more information that makes the problem solvable. We can assume that birds are not sold in halves or quarters, or in negative amounts. (And let's assume that you need to buy at least one of each animal too.) The values of x, y and z, therefore, must be positive whole numbers and, of course, less than 100.

Let's get to work. Multiply (2) by four, to get rid of the fraction:

(3) $20x + 16y + z = 400$

So, by some 'restoration':

$z = 400 - 20x - 16y$

If we substitute this for the z in (1) we get:

(4) $x + y + 400 - 20x - 16y = 100$

Which rearranges to:

$19x + 15y = 300$

We're down to a single equation with two variables, which we can solve because of the other constraints. The only positive, whole number values of x and y that work are $x = 15$ and $y = 1$, which you get through simple trial and error. (Note that 300 is divisible by 5. So $19x + 15y$ is divisible by 5. And since $15y$ is also divisible by 5, then $19x$ must also be divisible by 5, so x must be a multiple of 5. The only possible cases are $x = 5$, 10 and 15, and when you plug in the first two there are no solutions.) So, $z = 100 - x - y = 100 - 16 = 84$.

The answer is 15 cocks, 1 hen and 84 chicks.

In his treatise, Abu Kamil writes that, depending on the respective prices of the three birds, sometimes there is a single solution, as here, but that sometimes there is no solution, and sometimes several solutions. As an example he gives the following problem.

(52)

ONE HUNDRED BIRDS

If ducks cost 2 drachmae each, doves are two to the drachma and hens are three to the drachma, how many ducks, doves and hens do you have when you buy 100 birds with 100 drachmae? Find the six solutions.

As well as inventing new mathematics, medieval Arab scholars adopted the Indian number system of ten digits including a zero. 'Arabic' numerals – the digits 1, 2, 3, 4, 5, 6, 7, 8, 9 and 0 – reached Europe around the thirteenth century. One of the first European books to use Arabic numerals was the *Liber Abaci*, by Leonardo of Pisa. It contains lessons on calculation and measurement, as well as arithmetical puzzles including bird problems, such as this nice one with a unique solution: buy 30 birds for 30 denarii, with partridges at 3 denarii, pigeons at 2 denarii and sparrows at $\frac{1}{2}$ a denarius. (I'll leave you to solve it on your own.)

Over the following three centuries pretty much every top-ranking Renaissance mathematician produced their own bird problems, featuring bargains on the sale of thrushes, larks, blackbirds, flycatchers, capons, starlings, geese and many other feathery friends. Not only did they offer much entertainment as recreational puzzles, they also provided a cultural history of southern European ornithology (and gastronomy).

Once you learn how to solve a bird problem, you can solve them all: you rewrite the question as simultaneous equations and find the whole number solutions.

Many other puzzles also rely on turning the situation into simultaneous equations. Usually they don't have enough equations for the variables, so you have to introduce some sagacious trial and error, or mathematical insight, to solve them. This next one is my absolute favourite, not only because the amount of information given seems preposterously meagre – only two equations for *four* variables – but also because the number that is used is so fabulously on-brand.

(53)

THE 7-ELEVEN

A customer walks into a 7-Eleven shop and buys four items.
   'That's £7.11,' remarks the cashier.
   'How funny,' replies the customer.

'Yes,' says the cashier, 'all I did was multiply the prices of the four items together.'

'Aren't you supposed to add the prices together?'

'That's fine with me, when you add them the total is exactly the same.'

What was the price of each item?

To solve this problem, you will need to know a couple of simple maths facts. First, a prime number is a whole number divisible only by itself and 1. The list of prime numbers begins:

2, 3, 5, 7, 11, 13, 17, 19, ...

Second, you will need to know the fundamental theorem of arithmetic, the most important and basic rule about prime numbers, which states that every whole number can be written out as a multiplication of a unique set of prime numbers. For example:

$60 = 2 \times 2 \times 3 \times 5$

$711 = 3 \times 3 \times 79$

$123,456 = 2 \times 2 \times 2 \times 2 \times 2 \times 2 \times 3 \times 643$

In each case the number can be broken down in only one way into a multiplication of prime numbers. You may have taken this rule for granted without even knowing what it was called.

Anyway, the fundamental theorem of arithmetic will help you compose one of the equations necessary to solve this puzzle.

You may need to use a calculator, or a computer, to help you break down big numbers into primes. But it remains a fantastic problem even so.

What links the great nineteenth-century mathematician Siméon Denis Poisson and Bruce Willis, Hollywood action hero? They both cracked the following puzzle. In fact, Poisson's biographer wrote that the problem was the spark that turned the young Frenchman on to mathematics. 'Without ever having stopped to think about this sort of thing, without knowing the notation nor the methods of algebra, without ever having done any preliminary course, he solved [it] himself, and on that day he sensed the love of mathematics was born in him, that he should never give it up, and it was the origin of his glory.' Santé!

For Bruce Willis, the puzzle was equally life-affirming. In *Die Hard: With a Vengeance*, he and Samuel L. Jackson solve it in order to deactivate a time bomb. If Willis and Jackson can do it, you can too.

(54)

THE THREE JUGS

You have an 8-litre jug full of wine. You also have a 5-litre jug and a 3-litre jug, both empty. None of the jugs has any measurements on it.
Fill one of the jugs with exactly 4 litres.

The earliest appearance of this puzzle is in a thirteenth-century chronicle of the world written by Albert, an abbot from Stade, near Hamburg. The opus includes the most detailed medieval description of the pilgrim trail from northern Europe to Rome, which is written as a dialogue between two friars, Tirri and Firri. The men's merry banter includes several mathematical puzzles. 'Divide the wine,' says Tirri to Firri when he taunts him with the three jugs problem, 'Or go without.'

The puzzle is really fun to solve, and I'll leave you to do it the standard way, which is to pour between jugs and see where you end up. Do it before we move on.

Now I'm going to show you the *other* way to solve the three jugs puzzle, using balls bouncing around an unconventionally shaped billiard table.

The billiard table illustrated below is a rhomboid five units along one side and three units along the other, made up of equilateral triangles. I've shown the triangles, because they provide us with a coordinate system. The coordinate (x, y) is x along horizontally, and y up diagonally.

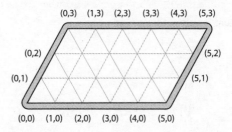

The following image shows what happens when you place a ball on position $(5, 0)$ and shoot it along the line of the triangle. It rebounds at $(2, 3)$, $(2, 0)$, $(0, 2)$, $(5, 2)$ and $(4, 3)$, before carrying on. (Mathematical pool tables have no friction and balls bounce perfectly the way they are supposed to.)

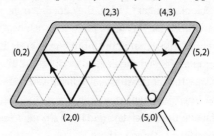

Now consider shooting a ball positioned at (0, 3). It rebounds at (3,0), (3, 3), (5, 1), (0, 1), (1, 0), (1, 3) and (4, 0), before continuing on its path.

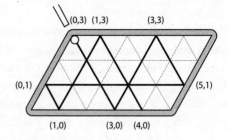

Let's look at those coordinates more closely:

SHOT 1	SHOT 2
(5, 0)	(0, 3)
(2, 3)	(3, 0)
(2, 0)	(3, 3)
(0, 2)	(5, 1)
(5, 2)	(0, 1)
(4, 3)	(1, 0)
	(1, 3)
	(4, 0)

Do these numbers look familiar? I hope they do! Since these are precisely the two possible solutions for The Three Jugs.

So we don't get confused, rename the 5-litre jug A, and the 3-litre jug B.

A and B start empty.

Let's fill A up. The state of the jugs is A = 5 litres and B = 0 litres. Write this as (5, 0).

Now pour A into B. A has 2 litres left, and B is full with 3 litres. The current state of the jugs is (2, 3).

Empty B into the third jug. The jugs are now on (2, 0).

Pour A into B: (0, 2)

Fill up A again: (5, 2)

Pour A into B: (4, 3)

A has 4 litres so we are done.

The volumes of A and B are exactly the coordinates of the rebounds when the ball is shot from position (5, 0).

If we had started the puzzle by filling up B first, the volumes of A and B would be described by the rebounds when the ball is shot from (0, 3).

The marvellous method for solving jug puzzles by bouncing balls around a billiard table was discovered by the British statistician M. C. K. Tweedie in 1939, when he was twenty years old. Every time a ball bounces on a rhomboidal table, its new trajectory takes you to the next move in the solution.

If you ever find yourself required to pour out a specific volume of liquid with a full container and two smaller, empty, unmarked containers of size X and Y – Willis and Jackson, if you're reading, take note – all you need to do is build an X by Y rhomboidal billiard table and shoot some balls.

THE TWO BUCKETS

You are by a stream with a 7-gallon and a 5-gallon bucket. How do you measure out 6 gallons in the fewest possible pourings?

The idea of pouring liquids between vessels gives rise to some other lovely brainteasers.

(56)

THE WHITE COFFEE PROBLEM

A flask contains black coffee. A bowl contains milk. You pour some coffee into the milk and then some of the cloudy mixture back into the flask so that the levels of the drinks in both containers are what they were at the start.

Is there more coffee in the bowl than there is milk in the flask?

That problem was for breakfast; the next puzzle is for later in the day.

(57)

WATER AND WINE

You have a pint of water in one jug and a pint of wine in another. Pour half a pint of the water into the wine. Stir. The wine jug now has a pint and a half of a water–wine mix. Pour half a pint of this mixture back into the water jug, so that both jugs contain a pint of liquid each. Stir. Continue transferring half pints between the two jugs.

After how many transfers will the percentage of wine be equal in each jug?

Liquid is not the only substance that pours. So does sand. Here's a version of the jugs puzzle where you must measure time, rather than volume.

(58)

FAMOUS FOR 15 MINUTES

Using only a 7-minute hourglass and an 11-minute one, time
a quarter of an hour exactly.

I'll give you a start with this one. We have two hourglasses, so we must start
by overturning both of them. If we overturned just one of them, we would
be able to measure to only 7 or 11 minutes, at which point we would be
back where we started.

Now look at the numbers. They are helpful to us. The hourglasses measure
7 and 11 minutes, and we need to measure 15. The difference between 7
and 11 is 4 – which is the same as the difference between 11 and 15, which
suggests the following strategy.

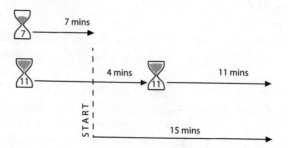

Overturn both hourglasses. Seven minutes later, the final grains of sand fall
in the 7 glass, so we know that the 11 glass has 4 minutes left to go. This
interval is just what we want, so let's begin our 15-minute period here. After
4 minutes the 11 glass is empty. Flip it over immediately and, 11 minutes
later, we have measured our 15-minute period exactly, as illustrated above.

This answer, however, is not the best one, since we required 22 minutes
in total to time our quarter of an hour. Find the better way.

Another way of measuring time is to light a fuse.

(59)

A FUSE TO CONFUSE

You have a set of fuses that each burn out in an hour. The fuses are long and thin, and burn unevenly, so one section may burn faster than another section. If you cut one in half there is no guarantee that each half will burn for half an hour.

[1] Using two fuses, measure 45 minutes.

[2] Using a single fuse, measure 20 minutes as accurately as possible.

The lesson with fuses that burn unevenly is that through mathematical insight we can eliminate the unevenness of burning and use them to time exact intervals. I love how the mathematics trumps the physics in this way.

Here's another puzzle about how to override material imperfections.

THE BIASED COIN

When you flip a fair coin the chance of getting heads or
tails is 50/50. Imagine you have an imperfect coin, where
the chance of getting heads or tails is not 50/50 but some
other fixed ratio. Is it possible to use the biased coin in such
a way that it can simulate a fair coin? You are looking for
some combination of biased flips that produces a 50/50
outcome.

Coins are crucial pieces of inventory in the puzzle universe, and we will
return to them in more detail in the next chapter.

The balancing scale was the only way to weigh objects until the eighteenth
century, when single-pan scales using springs were invented. A ubiquitous
measuring tool, the balancing scale was an obvious source of mathematical
puzzles from the Renaissance through to the Enlightenment, and beyond.
   Rattle the pans with this one.

DIVIDE THE FLOUR

You have a balancing scale and two weights, 10g and 40g.
Divide 1kg of flour into two piles, one of 200g and one of
800g, by balancing the scales three times only.

Let's say we have the following set of kilogram weights, made from the first six terms in the doubling sequence:

1, 2, 4, 8, 16, 32

We can combine these six weights to make every single kilogram weight between 1 and 63. For example:

3 = 2 + 1 (i.e., to make 3kg we add the 2kg to the 1kg)

13 = 8 + 4 + 1

27 = 16 + 8 + 2 + 1

63 = 32 + 16 + 8 + 4 + 2 + 1

In fact, the six weights above comprise the *smallest set of weights* that can measure every kilogram weight between 1 and 63.

(We can see why this is so by considering the weights as an expression of binary numbers. Binary is the counting system that uses only the digits 1 and 0. The numbers used in binary are those in our decimal system in which only 1 and 0 appear: 1, 10, 11, 100, 110, and so on. The numbers 1, 10, 100, 1000, 10000 and 100000 in binary correspond to the decimal numbers 1, 2, 4, 8, 16 and 32. Thus the binary numbers are instructions for how to build numbers using the doubling sequence.

3 in binary is 11

13 in binary is 1101

27 in binary is 11011

63 in binary is 111111

A 1 in the rightmost column is a 1, a 1 in the column next to it is a 2, a 1 in the column next to that one is a 4, and so on. Likewise, a 0 in the rightmost column means no 1s, a 0 in the column next to it means no 2s, a 0 in the column next to that one means no 4s, and so on. So, take 13, which is 1101. This represents, from right to left, one 1, no 2, one 4 and one 8. In other words, 13 = 1 + 4 + 8, as we saw above.)

Enough digression about binary numbers, fascinating though they are. Back to weights and scales.

Since our set of weights {1, 2, 4, 8, 16, 32} can measure out every kilogram from 1 to 63, we can weigh any whole number of kilograms from 1 to 63 by placing combinations of these weights on *one* of the pans of a balancing scale.

But what happens if you're allowed to use both pans?

BACHET'S WEIGHT PROBLEM

You have a set of balancing scales. What is the smallest set of weights that will weigh every whole number value from 1 to 40 kilograms when weights can be put on either pan?

This problem appears in Leonardo of Pisa's 1202 *Liber Abaci*, even though it is more commonly known as the weight problem of Frenchman Claude-Gaspard Bachet – and not because he liked extra frites with his cassoulet.

Bachet invented the puzzle book. A poet, translator and mathematician, in 1612 he published the first edition of *Problèmes Plaisants et Délectables Qui Se Font Par Les Nombres* (*Amusing and Entertaining Problems That Can Be Had With Numbers*). In it he gathered together many puzzles we've already seen in this book, such as the ones about crossing a river in a small boat, buying a hundred birds and pouring between three jugs. For three centuries *Problèmes Plaisants* was the standard text of recreational mathematics, and all subsequent puzzle literature is indebted to it. Bachet's book also includes the most famous analysis of the balancing scales problem.

Bachet's other pivotal role in the history of mathematics was as the Latin translator of the Greek text *Arithmetica* by Diophantus. The French mathematician Pierre de Fermat was reading Bachet's translation when he scribbled on one of its pages that he had a marvellous proof of a theorem inspired by the text, but that he could not write it down since the margin was too narrow to contain it. Fermat's Last Theorem – which states that there are no whole numbers a, b and c that satisfy the equation $a^n + b^n = c^n$ when n is bigger than 2 – eluded proof for more than 350 years, during which time it became the most celebrated unsolved problem in mathematics.

A preliminary puzzle:

> *You have eight identical coins. A ninth coin is counterfeit: it looks the same but is slightly lighter than all the others. Can you identify the counterfeit in only two weighings?*

You might want to solve this problem yourself, in which case look away now. I'm going to solve it for you here since it provides a leg-up to what's coming.

Solution:

Separate the coins into three groups of three. If we number the coins 1, 2, 3, 4, 5, 6, 7, 8 and 9, then our first weighing is 1, 2, 3 versus 4, 5, 6. Either the pans balance, or they don't.

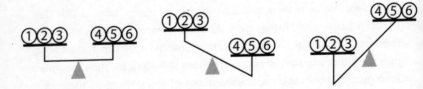

If they balance, as shown above left, we know that the lighter coin is one of 7, 8 or 9. If they tip as shown above middle, we know that the light coin is 1, 2 or 3; and if they tip as shown above right, the light coin is either 4, 5 or 6.

In all three cases, in other words, we are able to narrow the odds of selecting the light coin from 1 in 9 to 1 in 3.

For the second weighing, all we do is weigh one of the remaining coins against another, with the third left aside. The pan that tips up contains the lighter coin, or if the pans balance the lighter coin is the one left aside. *Voilà*.

The following puzzle went viral during the Second World War. It distracted Allied brains so much that it was suggested that the counterfeit coin be dropped on enemy territory to cause havoc with German brains too.

(63)

THE COUNTERFEIT COIN

You have eleven identical coins. A twelfth coin is counterfeit: it looks the same but has a different weight. You don't know if the fake coin is lighter or heavier than the others.

In three weighings of a balancing scale can you identify the counterfeit and determine whether it is lighter or heavier?

Single-pan scales – like the digital ones we use today, in which there is a single weighing pan that gives you a reading in kilograms – also provide clever puzzles involving counterfeit currency.

(64)

THE FAKE STACK

You have ten stacks of coins, with each stack containing ten pound coins. Nine of the stacks are made up of genuine pound coins, but a single stack contains only fake ones. You know the weight of a pound coin, and you also know that each fake coin weighs 1g more than a genuine pound coin. What is the smallest number of weighings required to identify the fake stack using a single-pan scale that gives you a reading of the weight on the pan?

The successor of Claude-Gaspard Bachet as the doyen of French number fun was Édouard Lucas, whose volumes on recreational maths appeared towards the end of the nineteenth century. A significant mathematician in his own right, making important advances in the understanding of prime numbers, Lucas invented new puzzles as well as analyzing the classics. The following story about Lucas is *absolument authentique*, according to a 1915

French maths text book. It took place many years ago, the author writes, at a scientific conference. Several well-known mathematicians – and a few illustrious ones – were milling around after lunch. Lucas piped up and challenged them to the puzzle below. A few replied with the wrong answer. Most were silent. No one got it right.

À vous, cher lecteur.

FROM LE HAVRE TO NEW YORK

Every day at noon in Le Havre an ocean liner sails to New York, and (simultaneously) in New York an ocean liner sails to Le Havre. The crossing takes exactly seven days and seven nights in either direction. How many other ocean liners will an ocean liner leaving Le Havre today pass at sea by the time it arrives in New York?

I love this puzzle because it takes a quotidian event – ships leaving and arriving in port – and finds a delicious mathematical morsel within it. When it comes to transport there are many great puzzles, which are often the kind of thing one thinks about when travelling.

(66)

THE ROUND TRIP

A plane travels from A to B and back again. If there is no wind the flight takes exactly the same amount of time both ways. But what happens if the wind is blowing? Does the round trip take more time, less time, exactly the same time, or does it depend on which way the wind is blowing?

We can assume that the wind is blowing in a constant direction for the entirety of the journey. Obviously, if you have a tailwind from A to B, a sudden change of wind direction, and a tailwind on the way back, the round trip will be faster than if there was no wind at all. And we can assume that the plane travels in a straight line from A to B, and in a straight line back again. Start by considering what happens if the plane travels in the direction of a tailwind on the way out – making that hop faster – and returns directly into a headwind – making the hop back slower. Does the effect of the wind cancel out overall or not? Follow on by thinking about about wind direction that is at an angle to the plane's direction.

Staring at the dashboard on a long drive can also provide arithmetical amusement.

(67)

THE MILEAGE PROBLEM

Modern cars usually contain two odometers. One tallies the total number of miles travelled in the car's life and cannot be reset, and the other (the trip meter) tallies the number of miles in a trip, and can be reset to zero. If either meter gets to the point where each of the digits is a 9, the next number that meter will display will contain only 0s.

Imagine that the first four digits of the odometer and the trip meter are the same, as below:

If you don't reset the trip meter, at what mileage on the odometer are the first four digits once more the same on both meters?

Now let's consider moving under our own steam.

(68)

THE OVERTAKE

[1] You are in a race. You overtake the person in second place. What position are you in?

[2] You are in a race. You overtake the person in last place. What position are you now in?

(69)

THE RUNNING STYLES

Constance and Daphne are running a marathon, which
we take as being a race of exactly 26.2 miles. Constance
runs the entire marathon at a constant speed of 8 minutes
per mile. Daphne runs at different speeds, with fast spurts
and slower sections, such that she covers every mile in 8
minutes and one second. In other words, whichever mile
you take – the first mile of the course, the last mile or, say,
the interval between 13.6 miles and 14.6 miles – Constance
will run it in 8 minutes and Daphne will be a second slower.

Is it possible that Daphne will win the race?

Without giving too much away, it is possible, and the trick is trying to find
the correct strategy. For me this question comes under the category of
paradox rather than puzzle. There are logical paradoxes – premises that
lead to self-contradictory conclusions – and there are cheekier paradoxes
– statements that suggest absurdity but which upon investigation are
actually true. Here are two puzzles in this vein.

(70)

THE SHRIVELLED SPUDS

A pile of potatoes weighing 100kg is put in the sun. Ninety-
nine per cent of the weight of the potatoes is made up of
water. After a day some of the water evaporates, with the
result that 98 per cent of the weight of the potatoes is now
made up of water. What's the new weight of the potatoes?

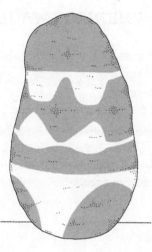

The next problem dates from the 1896 edition of W. W. Rouse Ball's *Mathematical Recreations and Essays*, the first important work of mathematical amusements written in English. The book, originally published in 1892, went on to have fourteen editions, the last four of which, published posthumously in 1939, 1942, 1974 and 1987, included revisions and extra sections by H. S. M. Coxeter, the great Canadian geometer. Among Rouse Ball's many activities at Cambridge, where he spent his academic career, was founding the Pentacle Club, one of the world's oldest magic societies. In his will he left money to both Oxford and Cambridge universities, which each set up a Rouse Ball professorship of mathematics.

(71)

THE WAGE WAGER

You are offered a new job with a starting salary of £10,000 a year. Your boss asks you to choose one of two ways that your salary can grow.

[1] Plan A: a £500 raise every six months (meaning that every six months your salary for the next six months rises by £500).

[2] Plan B: a £2,000 raise every year.

Which plan do you choose?

(72)

A STICKY PROBLEM

Dick has a stick. He saws it in two. If the cut is made at random anywhere along the stick, what is the length, on average, of the smaller part?

One of Édouard Lucas's best-known puzzles is the *ménage*, or couples, problem, which asks how many ways there are to seat male/female couples at a table such that men and women sit in alternate positions and no man sits next to his wife. The solution is far too complicated for this book, and Lucas included it as an appendix to his academic book on number theory, rather than in his books of mathematical recreations. But, since you asked, with two couples it is impossible, with three couples there are 12 arrangements, with four 96 and with five 3,120.

The idea that dinner parties make for great puzzles, however, is an inspired one.

(73)

THE HANDSHAKES

Edward and Lucy invite four couples to dinner. Each person shakes hands only with the people that he or she has not met before. Edward then asks his wife and eight guests how many hands they shook, and he receives nine different answers.
    How many people did Lucy shake hands with?

If that was a bit formal, we can make it less so.

(74)

THE HANDSHAKES AND THE KISSES

Edward and Lucy invite some friends for dinner. Some are single and some are in opposite-sex couples. Men greet each other with a handshake. Women greet both men and women with a kiss. (And, of course, the two people in any couple do not greet each other.) At the party, each guest greets both Edward and Lucy and all the other guests. Given that there were 6 handshakes and 12 kisses in total, how many people came to the dinner party and how many singletons were there?

Dinner party puzzles are about working out combinations. Sometimes there seem almost too many combinations to count. So don't count them. And when you go to the theatre, don't forget your ticket.

(75)

THE LOST TICKET

One hundred people are lined up to take their seats in a 100-seat theatre. The first in line can't find her ticket so she sits on a seat at random. Each remaining theatregoer sits in his or her assigned seat, unless it is occupied, in which case he or she sits on a seat at random.

What is the probability that the last person in the theatre takes their assigned seat?

This chapter contained puzzles about hypothetical real-world situations. Now it's time to get physical.

Ten tasty teasers

Are you a genius at geography?

1) What is the largest city in Europe whose name (in English) has only one syllable?

Prague   rome

2) Which U.S. state is closest to Africa?

Florida
North Carolina
New York
Massachusetts
Maine

3) Order the following cities from west to east.

Edinburgh
Glasgow
Liverpool
Manchester
Plymouth

4) Order the following cities from north to south.

Algiers
Halifax, Nova Scotia
Paris
Seattle
Tokyo

5) Order the following cities from north to south.

Buenos Aires
Cape Town
Easter Island
Montevideo
Perth, Australia

6) Which European country shares a border with the most European countries?

7) List the following in order of population.

Shetland Islands
Isle of Man
Isle of Wight
Jersey
Falkland Islands

8) Which country in the world has the longest coastline?

9) France is the country with the largest number of time zones (12), because the French Republic includes its overseas territories and departments. But which is the largest country to have a single time zone?

10) Aconcagua, Mount Elbrus, Mount Kilimanjaro and Mount McKinley are the highest peaks in South America, Europe, Africa and North America. Rank them in order of height.

Your Aid I Want, Nine Trees to Plant

PROBLEMS WITH PROPS

Puzzles that involve playing around with actual objects can be the most engrossing of all mathematical recreations. We disappear into deep concentration much more effortlessly when our attempts to solve a puzzle are unmediated by scribbling on paper or thinking in abstractions. Not only is it enthralling to physically connect with a puzzle – to take it on *mano a mano* – fiddling around with real things makes the whole experience much more like playing with a toy or a game.

In this chapter we'll be entertaining ourselves with coins, matches, stamps, paper and string. The contents of your pocket or purse. The first problem requires only four identical coins. If you have ever shared a train journey with me, I will have undoubtedly challenged you with it.

> *Can you arrange four identical coins on a table in the pattern shown below, so that a fifth coin can be slid without hindrance into the shaded position that touches all the coins?*

The question is asking you to find a method to place the four coins in this way that guarantees that the fifth coin will fit between them perfectly. Try it. You cannot put them in that position by eye alone.

The solution is as follows: first you need to work out how you can fix the relative position of four coins. There's only one way to do this, which is to arrange them in a rhombus shape, as shown in Step 1 opposite. To make sure the coins remain in fixed relative positions, when any coin is slid from its position it must be placed in a new position in which it is touching two other coins.

The problem becomes, therefore, about getting from the arrangement shown in Step 1 to the arrangement shown in the question above, with the golden rule that each move must put a coin into a position in which it is touching two other coins. The solution is shown in Steps 2 and 3.

Step 1   Step 2   Step 3

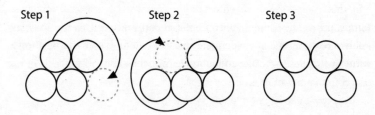

Coin puzzles offer fantastic bite-sized challenges. They are utterly absorbing until – pardon the pun – the penny drops. The solutions are generally much trickier than you think they will be, but you can usually find them with some thought.

The problem we've just completed was devised by Henry Ernest Dudeney a century ago, and the next one is also one of his.

(76)

THE SIX COINS

Can you arrange six identical coins on a table in the pattern below so that a seventh coin would fit into the shaded position that touches all the coins?

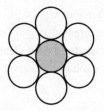

First you need to come up with an initial pattern that fixes the six coins' relative positions. Each move then consists of sliding a coin to a position in which it touches two other coins. You cannot lift a coin off the table or use one coin to push another.

The puzzle can be completed in three moves.

Coin puzzles are especially addictive. Once you have cracked the one above, you will be craving another.

77

TRIANGLE TO LINE

Can you go from the triangle to the line in seven moves? As above, each move consists of sliding a coin to a position in which it touches two other coins. You cannot lift a coin off the table or use one coin to push another.

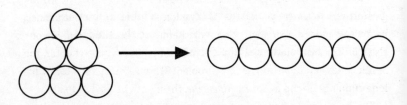

For the following challenge you need eight identical coins.

(78)

THE WATER PUZZLE

Using the moving rules of the previous questions, can you go from H to O in four moves?

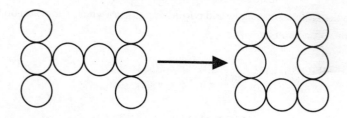

And can you return from O to H in six?

We've met Henry Ernest Dudeney once before in this book. He was the creator of the Smith, Jones and Robinson puzzle that sparked a boom in inferential logic problems in the 1930s.

Dudeney was Britain's greatest puzzle innovator, and arguably the world's greatest too. In a forty-year career writing for newspapers and magazines, he devised more classic recreational mathematics problems than anyone else. His creativity never stopped. As mentioned in the first chapter, the Smith, Jones and Robinson puzzle (Problem 7) appeared in *Perplexities*, his long-running column in *Strand Magazine*, the month he died in 1930.

The roots of Dudeney's vocation can perhaps be traced back to his grandfather, a shepherd on the South Downs who taught himself maths and astronomy while herding his sheep. He eventually became a schoolteacher, as did his son. Dudeney was born in 1857, but he was not one for institutionalised education. Rather than stay at school, at the age of

13 he started work as a clerk in the civil service in London. The salaryman life eventually bored him and he began to submit puzzles to the national press, before leaving to produce puzzles full time.

Dudeney's output is remarkable not only for its breadth but also for its depth, which is especially notable for an autodidact. His arithmetical agility was off the scale. In his first book, *The Canterbury Puzzles* (1907), one question notes that the cubes of one and two make nine ($1^3 + 2^3 = 1 + 8 = 9$). He then asks the reader to find two other numbers whose cubes add up to nine. The answer is:

$$\frac{415,280,564,497}{348,671,682,660} \quad \text{and} \quad \frac{676,702,467,503}{348,671,682,660}$$

'An eminent actuary and another correspondent have taken the trouble to cube out these numbers, and they both find my result quite correct,' Dudeney wrote. The mind self-destructs in awe thinking about how he found this result using only paper and pencil.

Dudeney devised many coin puzzles. This next one is taken from *Amusements in Mathematics* (1917).

(79)

THE FIVE PENNIES

Here is a really hard puzzle, and yet its conditions are so absurdly simple. Every reader knows how to place four pennies so that they are equidistant from each other. All you have to do is to arrange three of them flat on the table so that they touch one another in the form of a triangle, and lay the fourth penny on top in the centre. Then, as every penny touches every other penny, they are all at equal distances from one another. Now try to do the same thing with five pennies – place them so that every penny shall touch every other penny – and you will find it a different matter altogether.

This puzzle is a little less tricky – or at least it is for someone as clumsy as me – if you try it with the largest coins you can find, like two pence or ten pence pieces. Dudeney only presented one solution, but there are two.

Rational Amusement for Winter Evenings, a book from 1821 by John Jackson, contains the following verse:

> *'Your aid I want, nine trees to plant*
> *In rows just half a score;*
> *And let there be in each row three.*
> *Solve this: I ask no more.'*

I'll translate: can you plant nine trees in ten rows, with three trees per row? Here's the solution.

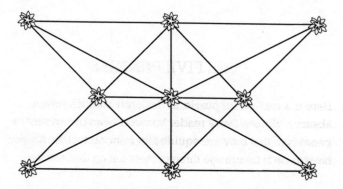

In *Amusements in Mathematics*, Dudeney writes that this problem is attributed to Isaac Newton, although no evidence exists that it pre-dates Jackson. Tree-planting puzzles – in which you must arrange a certain number of trees in a certain number of lines containing a certain number of trees – are most conveniently tackled using coins. The next puzzle is one of Dudeney's.

80

PLANTING TEN TREES

Place ten coins on a large piece of paper, as shown below.

Can you remove four coins and replace them on the paper so that the ten coins now create five straight lines with four coins on each line?

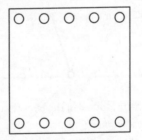

If you manage to do this – and Dudeney said it was not difficult – can you work out how many different ways there are to solve this puzzle, assuming you start from the initial position each time?

This puzzle can get confusing if the coins shift slightly, so I'd recommend marking the initial positions on paper.

Dudeney wrote several problems involving ten points, or 'trees', in five lines of four. If you disregard the rule in the last puzzle that you can only move four coins – so that you can move as many as you like – there are five other arrangements of ten coins that produce five straight lines with four coins on each line. Dudeney called these arrangements the 'star', the 'dart', the 'compasses', the 'funnel' and the 'nail'. (He called the solution to the previous question the 'scissors'.) The star is shown below. Can you find the other four?

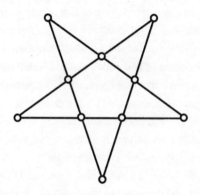

Here's a riddle: why at the beginning of the twentieth century was the most famous Henry Dudeney a woman?

Alice Whiffin was eighteen when she married Henry Ernest Dudeney. She later became a successful novelist, favourably compared to Thomas Hardy and Edith Wharton, and she chose to use her married name, Mrs Henry Dudeney, on her books. She wrote fifty or so novels, many set in the provincial south-east. The Dudeneys built their own country estate and were fixtures in London's literary set. Alice was a confidante of Sir Philip Sassoon, the extremely wealthy MP and socialite known for hosting starry parties at his Kent mansion.

The Dudeneys also had a tempestuous relationship. She had an affair. They separated. In 1916, they moved back in together in a house in the town of Lewes, where they had a study each, one below the other. Alice wrote a detailed diary of her time with Ernest, as she called him, in Lewes. The diary was eventually published in 1998 and provides an affectionate if spiky portrait of life with the world's greatest puzzle inventor. At times it is almost uncomfortably intimate. He is adoring of her but prone to jealous rages. 'Ernest's temper is abominable, but he can't help it, doesn't even know it (that's my belief) and makes exquisite amends,' she wrote. 'After all, if you are married to a genius – and are by way of being one yourself – [there are] bound to be clashes ...'

One extraordinary talent of Dudeney's was his ability to take an everyday object and turn it into a mathematical puzzle. He came up with the following ingenious problem, about cigars, at his London club. 'For a considerable period it absorbed the attention of the members,' he wrote. 'They could make nothing of it, and considered it quite impossible of solution. And yet, as I shall show, the answer is remarkably simple.'

The puzzle works for cigars, but I think it's easier to find the solution if you consider coins, so I've rewritten the question. (Replace every mention of 'coin' with 'cigar' if you want to answer the original version.)

(81)

THE SPACE RACE

Two players are seated at a square table. The first player places a coin on the table, the second places a coin on the table, and they carry on placing coins one after another, with the only condition being that the coins are not allowed to touch. The winner is the person who places the final coin on the table, meaning that he or she fills the last remaining space between the other coins.

The table has to be larger than a single coin, otherwise the game ends on the first go, and all the coins placed must be identical.

One player is always guaranteed to win. Which one is it – the one who starts or the one who goes second? And what is his or her strategy?

We'll return to Dudeney later, but now that our coins are on the table I'd like to have some more fun with them.

In his 1883 introductory address to the Edinburgh Mathematical Society, the Scottish mathematical physicist Peter Guthrie Tait claimed to have spotted the following puzzle on a train. It's reassuring to me that coin puzzles have been distracting passengers for as long as there have been railways.

Tait's contributions to science were many and varied. In maths he founded the theory of knots. A keen experimenter and a prolific writer – he co-wrote a classic physics textbook with Lord Kelvin – he was adored by his students in Edinburgh for fantastic science demonstrations that featured giant magnets, water jets and electrical sparks that would fill the room. His favourite prop, however, was probably the golf club. His obsession with the game led to papers on the trajectories of golf balls, or 'rotating spherical projectiles'. His son Freddie became a well-known golfer, winning the British Amateur Championship twice.

The next puzzle is believed to be Japanese in origin, although it was Tait who popularised it in the West.

(82)

TAIT'S TEASER

Arrange two types of coin as shown below in 1, alternating between one and the other. Or, if you only have one type of coin, alternate between heads and tails. The aim is to change the positions of the eight coins so that the four of each type are adjacent, as shown in 2.

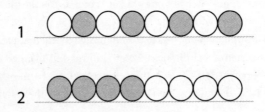

Each move consists of moving two adjacent coins at the same time. You can move them to wherever you like on the same line as the other coins, but the two moving coins cannot switch positions: during the move the one on the left must remain on the left, and the one on the right must remain on the right.

Can you complete the puzzle in four moves?

It's easy to get disheartened with this puzzle if you can't do it instantly, but persevere, because it can be done. To help you, here's how to solve a simpler version with six coins. Note that in the final position, all the coins have shifted to the left by two spaces.

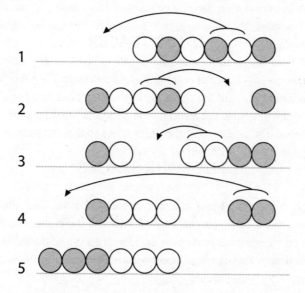

While we're on the subject, I thought I'd sneak in a final Tait-style problem. This one only has five coins, but comes with the extra condition that you may only move two *different* coins each move. Can you get the coins from the order in the top row overleaf to the order in the bottom row in four moves?

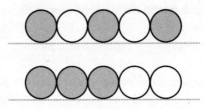

The nineteenth-century French mathematician Édouard Lucas, who we met last chapter challenging colleagues with a problem about ocean liners, included the following two traditional puzzles in *Récréations Mathématiques*.

They both belong to the category of utterly bewildering yet devilishly simple.

(83)

THE FOUR STACKS

Eight coins are placed in a row, as shown below. Each time you move a coin you can move it either to the right or to the left so it hops over two coins and lands on top of the third coin along. You can hop over two single coins or a stack of two.

Can you move four coins so you end up with four stacks of two coins?

Lucas called the next puzzle the *jeu des grenouilles*. He suggested using black and white pawns, but in the absence of a chess set, coins will do.

(84)

FROGS AND TOADS

Place three coins of one denomination and three coins of another denomination in a line, as shown below, with a single space in the middle. (Or use three heads and three tails.) On the left are the frogs and on the right the toads. Frogs can only move left to right, and toads only right to left. A frog or toad can either move one place forward to a vacant position, or it can hop over a single coin to a vacant space, so long as it is moving in the correct direction.

Can you move all the frogs into the positions of the toads, and the toads into the positions of the frogs?

Solitaire – or peg solitaire – is probably the best-known one-player game that involves pieces hopping over other pieces. 'The game called solitaire pleases me much,' wrote the great German polymath Gottfried Leibniz in 1716. Leibniz's contributions to science and philosophy included discovering calculus (independently of Isaac Newton), inventing calculating machines, and cheerleading the cause of binary numbers, whose 0s and 1s corresponded to the holes and pegs of his beloved solitaire. Leibniz, however, preferred playing the game in reverse: rather than jumping over a peg into a vacant space and removing the jumped-over peg, he would jump over an empty space and put a peg in it. 'But why all this? You ask,' he wrote. 'I reply: to perfect the art of invention.'

Here we're going to play coin solitaire. The normal rules apply: any coin can hop over an adjacent coin to the space on the other side, and then the coin that was hopped over is removed. As in draughts you can, if you like, hop over several coins in one move if the coin lands in a position where another hop is possible.

85

TRIANGLE SOLITAIRE

Place ten coins in a triangle, as shown below. Remove a single coin. Now, by hopping coins over other coins, reduce the grid to a single coin.

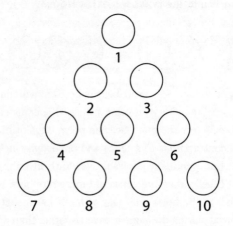

Like the previous coin puzzles, this one will also grab hold of you until you solve it. Before you try, I'd recommend placing the coins on a piece of paper with 10 marked dots so the coins don't lose their positions in the grid.

After playing around for a while you may discover the following solution in six moves after removing the coin from position 2:

1. The coin on 7 moves to 2. (4 is removed.)
2. 9 moves to 7. (8 is removed.)
3. 1 moves to 4. (2 is removed.)
4. 7 moves to 2. (4 is removed.)
5. 6 moves to 4 then to 1, and then to 6 . (5, 2 and 3 are removed.)
6. 10 moves to 3. (6 is removed.)

But we can do better. Find the five-move solution.

I've included half a dozen coin puzzles – almost half the chapter – because coins are the most versatile of all puzzle props. Their physical properties can be exploited in a large number of ways: they slide and stack like counters; they can be used as geometrical points or solitaire pegs. They also have two identifiably different sides, which forms the basis of the next puzzle, or magic trick.

(86)

COINS IN THE DARK

You, the magician, are blindfolded. You ask an audience member to put ten coins flat on the table in front of you, and to tell you how many of them are showing heads.

You cannot see the coins, nor can you tell which way up they are by touching them.

How do you separate the coins into two groups, such that each group has the same number of heads?

When I first saw this trick I was impressed – but not as impressed as when I saw the solution.

Attack the problem first without a blindfold. Get ten coins. Lay them out on the table with, say, three heads visible. Try to separate the coins into two groups such that both groups end up with the same number of heads. Since the three heads do not divide evenly into in two quantities of heads, you will need to flip some coins over. The key to solving this puzzle is deciding which ones to flip, and then working out how you can make this decision while wearing a blindfold.

The Coins in the Dark puzzle works as a magic trick because the 'aha!' is so pleasurably surprising. In fact, puzzles are often like magic tricks. Not just because a solution is a kind of reveal, but also because the question often relies on subtle misdirection.

I've included the next puzzle because it involves coins, and produces a fascinating result. I don't expect you to physically lay out one hundred coins, but I do recommend that you read the solution even if you cannot solve it. It was first set at the 1996 International Olympiad in Informatics, a competition for pre-university students, so was originally aimed at super-smart teenage computer geeks. It's a wonderful problem.

(87)
THE ONE HUNDRED COINS

A row of one hundred coins is laid out on a table. Penny and Bob are playing a game in which the aim is to gather the most money by picking the coins off the table one by one. They are only allowed to pick the coins that are at the ends of the row. The coins are of different values – some are one pence pieces, others two pence pieces, pound coins, and so on.

Penny starts. She takes a coin from one end and pockets it. Bob chooses a coin from one end and pockets it. They carry on taking turns until there are no coins left. On each turn they can choose to take a coin from either end.

Can you prove that Penny can always make at least as much as Bob?

Here's a hint. Number the coins from 1 to 100.

Okay, enough coins already! We're going to move on to what is historically the second most popular puzzle prop, the matchstick. But before we get there, here's a puzzle that includes both a coin *and* a matchstick. Consider it the recreational mathematics equivalent of a rare duet between two famous old crooners.

88

FREE THE COIN

Two upturned glasses are positioned as below. A match
rests between them and a coin is trapped in the left glass.
Can you remove the coin without letting the match fall?

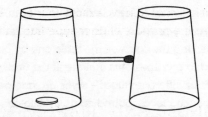

The safety match was invented in the mid-nineteenth century, and for a
hundred years matchstick problems were probably the most widespread,
cross-generational type of puzzle. Matchsticks are not as ubiquitous as
they once were, because fewer people smoke, and those that do smoke use
lighters. Toothpicks, pencils or cotton buds make for acceptable substitutes.

Henry Ernest Dudeney describes the following as 'a little puzzle for
young readers'.

89

PRUNING TRIANGLES

The 16 matches opposite form eight equilateral triangles.

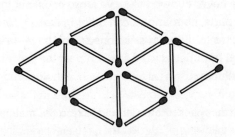

Remove four matches to leave exactly four equilateral triangles, leaving no loose ends or superfluous matches.

90

TRIANGLE, AND TRIANGLE AGAIN

Place 12 matches in a hexagon of six equilateral triangles.

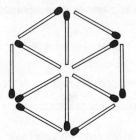

This puzzle has four parts.

[1] Move two matches to different positions so that *five* equilateral triangles are left.

[2] From this new pattern, move two matches to different positions so *four* equilateral triangles are left.

You are never allowed to leave loose ends, but in the next two parts the triangles can differ in size. [3] Move two matches to different positions so you have *three* equilateral triangles, and [4] finally repeat the process to be left with *two*.

Now let's go in the opposite direction and increase the number of triangles. I really like the following puzzle because it requires so few matches.

91

GROWING TRIANGLES

[1] Here are two triangles made from six matches. Can you move two matches to new positions so that there are four triangles? You are allowed to overlap matches.

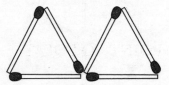

[2] Now make four triangles with six matches – without overlapping matches.

Earlier I asked how to place five coins so that each coin touches the others. Here's the match version of that puzzle.

(92)

A TOUCHING PROBLEM

Sticking with six matches, can you arrange them in such a way that each match touches every other match?
Now find a way to do the same with seven matches.

An arrangement of matches where they only touch each other at their endpoints can be understood in two ways. First, as an arrangement of matches. But we can also think of it as a network of points connected by matches, as in this next puzzle.

(93)

POINT TO POINT

Find an arrangement of 12 matches such that both ends of every match touch the ends of exactly two other matches.
In other words, create a network of points connected by matches such that every point links to three other points.

Let's finish our match puzzles with this twist on the topic by our old friend Henry Ernest Dudeney.

(94)

THE TWO ENCLOSURES

Shown below are 20 matches divided into two rectangular enclosures, comprising 6 and 14 matches respectively. The second rectangle has three times the area of the first.

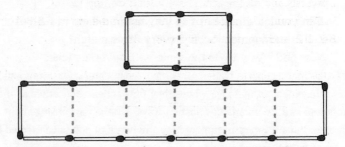

By transferring one match from the big enclosure to the smaller one, so you have one group of 7 matches and another of 13, can you create two new enclosures, where the second is still three times as big as the first?

Reading through his work, I am endlessly dazzled by Dudeney's ability to find brilliant material using random stuff you might find in your pocket. One exceptional puzzle is the following, which uses a block of eight stamps. If you don't have stamps, cut and fold a piece of paper.

It's time now to get a pair of scissors anyway, as you'll need them for the remaining puzzles.

FOLDING STAMPS

The illustration below shows a block of stamps in which the stamps are marked from 1 to 8. The aim of the puzzle is to fold the stamps along their borders so that the 1 is face upwards and all the others are folded behind it.

Can you fold the stamps so they are in the order 1-5-6-4-8-7-3-2, and (harder) 1-3-7-5-6-8-4-2?

'It's a fascinating puzzle,' Dudeney beamed. 'Do not give it up as impossible!'

1	2	3	4
5	6	7	8

Dudeney also came up with the following problem involving a block of square stamps.

(96)

THE FOUR STAMPS

You have a set of 12 square stamps in a 3 × 4 grid, as shown below, and a friend asks you for four of them. You decide to tear out four stamps that remain joined together. For example, 1-2-3-4, or 1-2-5-6, or 1-2-3-6 or 1-2-3-7, and so on. The stamps cannot be hanging together at a corner but can join on any side.

How many possible sets of four connected stamps are there?

1	2	3	4
5	6	7	8
9	10	11	12

In the answer to this puzzle at the end of the book I've drawn the possible shapes we can make from four connected stamps. Take a peek once you've solved it. Are they familiar?

Yes, Henry Ernest Dudeney's stamp puzzle features the family of shapes known as Tetris blocks.

For the small number of you who have never played it, Tetris is a very simple and utterly addictive computer game in which sets of four connected squares – the shapes on page 284 – fall from the top of the screen. You must then stack them on each other by moving them horizontally or rotating them.

The inventor of Tetris, Alexey Pajitnov, was inspired by the work of Solomon Golomb, a mathematician who published a book about shapes that are made by connected squares in 1965. Golomb was himself inspired by Dudeney.

Dudeney had no formal training, yet he had an amazing instinct for introducing ideas in his puzzles that mathematicians would later find worthy of academic study.

The Canterbury Puzzles, Dudeney's first book, contains his first puzzle to involve shapes made from connected squares.

The puzzle was inspired by a story in John Hayward's (not entirely reliable) 1613 history of William the Conqueror. Henry and Robert, William's sons, visit Louis, the heir to the French throne. When Henry beats Louis at chess a brawl ensues. 'Henry again stroke Louis with the chessboard, drew blood with the blowe,' wrote Hayward. 'Hereupon [Henry and Robert] presently went to horse, and their spurres claimed so good haste [...] albeit they were sharply pursued by the French.' Ooh la la.

THE BROKEN CHESSBOARD

A chessboard is broken into the 13 fragments shown below, which are all the possible shapes made from five connected squares, plus a four-square block.

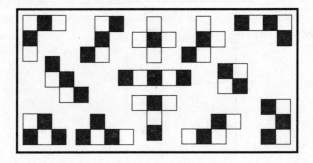

Can you reconstruct the chessboard?

Dudeney suggests that you cut the shapes out of squared paper and mount them on cardboard. 'They will form a source of perpetual amusement in the home,' he wrote. 'If you succeed ... but do not record the arrangement, you will find it just as puzzling the next time you feel disposed to attack it.'

Since we've got the paper and scissors out, here's a very enjoyable paper-folding puzzle which you will solve faster than the last one.

98

FOLDING A CUBE

Cut out the middle square from a grid of 3 × 3 squares, as shown below.

Can you fold the ring of eight squares to form a cube? Since a cube has six sides, two of the squares will have overlaps.

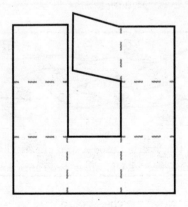

If you were ever in the Scouts or the Guides, and you learned how to make a woggle, your childhood was not wasted. That knowledge will (at last) prove useful.

(99)

THE IMPOSSIBLE BRAID

Cut a thin strip from a plastic bag and make two slits in it, as shown in A below.

Can you braid it so it looks like B?

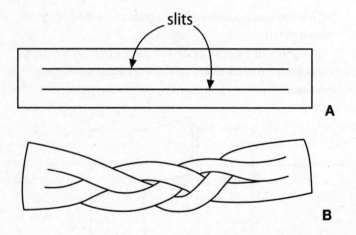

When I first tried this puzzle I used a strip of paper, but paper tears, so plastic is much more effective. Or leather, if you are a Scout or a Guide.

There is no thinking outside the box for how to braid the three strands: it's simply a matter of manipulating the strip in the right way. Note that the three strands pass around each other just as plaited hair does. They cross each other at six points, and they stay flat as they intertwine. Give it a whirl!

For the final puzzle in this chapter we need some more props: some string and some cardboard. Cut out two small rectangles of cardboard and join them with string as shown below. Write 'FRONT' on the face-up side of each piece of cardboard.

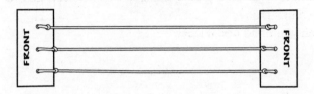

This model is topologically the same shape as the strip in the previous question: both comprise three strands joined at either end. We've made it out of string, however, since this allows us to explore some of its physical properties.

Piet Hein, a Danish poet and recreational mathematician, popularised the next puzzle after he learned about the string model during his frequent visits to Niels Bohr's Institute for Theoretical Physics in Copenhagen in the 1930s.

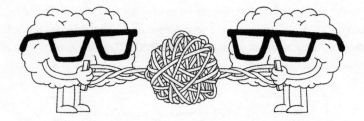

⑽⑽

TANGLOIDS

Hold the left end of the cardboard and string model, and rotate the right end between the top two strings for a full rotation, as illustrated opposite in A. In order for it to be a full rotation, the word 'FRONT' on the right piece of cardboard must again be facing up. The model will then look like B. Now fully rotate the right side between the bottom two strings, as shown, and the position of the strings should be as in C.

Can you untangle the string without rotating any of the pieces of cardboard?

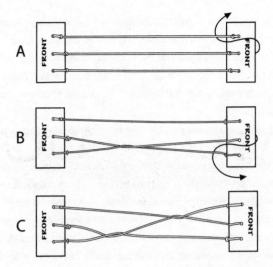

To make sure you don't rotate the pieces of cardboard, hold the left one in your left hand, and the right one in your right hand. Make sure that the word 'FRONT' is always facing you on both pieces and that the pieces are level with one another. Since you cannot rotate them, all you can do is slide the pieces of cardboard between the strings.

Keep on sliding the cardboard and the strings will disentangle. It feels miraculous, and I think this is possibly the most pleasurable puzzle in this chapter. What is more satisfying than the effortless unsnarling of knotted string?

To guarantee you this joy, I've decided not to print the solution in the answers section. You will have to solve it on your own.

Once you've done so, you will be so captivated by the model that you will want another disentanglement puzzle. In which case simply repeat the process in the question whereby the left piece of cardboard is held still and the right piece is rotated through *two* full rotations. Instead of the first rotation being between the top two strings from the front, you could rotate it between them from the back, or you could rotate it between the bottom two strings, either from the front or the back, or you could just twist it round as if turning a dial 360 degrees. The second rotation could be any of these too.

If you made only one rotation, it is impossible to disentangle the strings just by sliding the pieces of cardboard between them. But if you made two rotations, the tangle can be disentangled *whichever* rotations you chose.

Piet Hein suggested that the way to best enjoy this puzzle is to play it as a two-player game, which he called 'tangloids'. One player holds the left end of the model, and the other the right. The first player makes two rotations to his piece of cardboard and his or her opponent must untangle the mess. Players take turns in setting and solving, and the person who untangles fastest is the winner.

The fascinating property of the string model that double rotations can always be untangled but single rotations cannot helps explain the behaviour of certain rotations in space, which is why Niels Bohr and his colleagues were interested in it. The British quantum physicist Paul Dirac, who spent time in Copenhagen, used it as a teaching aid 'to illustrate the fact that the fundamental group of the group of rotations in 3-space has a single generator of the period 2'.

A good puzzle, as we saw earlier, can be like a magic trick. It can also be a brilliant way to explain serious science.

Ten tasty teasers

Are you smarter than a 13-year-old?

Class rules: no calculators allowed.

1) How many statements in the box are true?

> None of these statements are true.
> Exactly one of these statements is true.
> Exactly two of these statements are true.
> All of these statements are true.

A 0 **B** 1 **C** 2 **D** 3 **E** 4

2) Which one of the following shapes could not appear as the overlapping region of two identical squares?

A equilateral triangle
B square
C kite
D heptagon
E regular octagon

3) Exactly one of these equations is correct. Which one?

A $44^2 + 77^2 = 4477$
B $55^2 + 66^2 = 5566$
C $66^2 + 55^2 = 6655$
D $88^2 + 33^2 = 8833$
E $99^2 + 22^2 = 9922$

4) In how many different ways can a row of five 'on/off' switches be set so that no two adjacent switches are in the 'off' position?

A 5 **B** 10 **C** 11 **D** 13 **E** 15

5) In the addition shown below, each letter stands for a different digit, with S standing for 3. What is the value of Y x O?

$$\begin{array}{r} S\ O \\ +\ M\ A\ N\ Y \\ \hline S\ U\ M\ S \end{array}$$

A 0 **B** 2 **C** 36 **D** 40 **E** 42

6) On a digital clock that displays hours, minutes and seconds, how many times in each 24-hour period do all six digits change simultaneously?

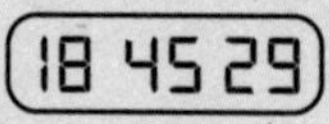

A 0 **B** 1 **C** 2 **D** 3 **E** 24

7) One of the following cubes is the smallest cube that can be written as the sum of three positive cubes. Which is it?

A 27   **B** 64   **C** 125   **D** 216   **E** 512

8) In a sequence of numbers, each term after the first three terms is the sum of the previous three terms. The first three terms are –3, 0, 2. What is the first term to exceed 100?

A 11th term   **B** 12th term   **C** 13th term   **D** 14th term   **E** 15th term

9) The pages of a book are numbered 1, 2, 3, … . In total it takes 852 digits to number all the pages of the book. What is the number of the last page?

A 215   **B** 314   **C** 320   **D** 329   **E** 422

10) The diagram below shows a unit cube (meaning its length, width and height are 1). Imagine it is coloured blue. Additional blue unit cubes are now glued face-to-face to each of its six faces to form a three-dimensional 'cross'. If unit cubes coloured yellow are now glued face-to-face to all the spare faces of this cross, how many yellow unit cubes are required?

A 6   **B** 18   **C** 24   **D** 30   **E** 36

The Number Games

PROBLEMS FOR PURISTS

A book of mathematical problems would not be complete without some number puzzles. Not puzzles that rely on numbers – many puzzles do that, as we have seen in previous chapters – but problems that unashamedly celebrate numbers and the patterns they reveal. These puzzles require neither props nor whimsy to sweeten the pill. They give it to you straight. Yet despite this honesty, number puzzles can be remarkably playful. There is fun to be had even in a procedure as simple as adding up.

Can you sum the numbers from 1 to 100?

This old chestnut was solved instantly by Carl Friedrich Gauss in the late eighteenth century when the great mathematician was still in short trousers. Or so the story goes. His teacher was expecting him to *actually* add the numbers up, one by one, but the boy genius spotted a pattern.

$1 + 2 + 3 + 4 + ... + 97 + 98 + 99 + 100$

is the same as adding pairs, one from the front and one from the back:

$(1 + 100) + (2 + 99) + (3 + 98) + (4 + 97) + ... + (50 + 51)$

Which are all the same:

$101 + 101 + 101 + 101 + ... + 101$

So the total sum is 101 fifty times, or $101 \times 50 = 5,050$

Clever Carl! The story is usually told as if he was the first person to have this idea. Yet the same problem was in Alcuin's *Propositiones ad Acuendos Juvenes* one thousand years earlier.

> *There was a ladder with 100 steps. One pigeon sat on the first step, two pigeons on the second, three pigeons on the third, four on the fourth, five on the fifth, and so on up to the hundredth step. How many pigeons in total were there on the ladder?*

The set-up is different, but the arithmetic is evidently the same as above – adding the numbers from 1 to 100. Alcuin's solution also involved adding pairs, but different ones. He took the first step of the ladder and the penultimate step together, to get $1 + 99 = 100$, and then the second and the antepenultimate step, and so on.

The sum is therefore:

$(1 + 99) + (2 + 98) + (3 + 97) + ... + (49 + 51)$ plus 50 from the fiftieth step and 100 from the hundredth step.

Which is:

$(49 \times 100) + 50 + 100 = 4900 + 150 = 5050$

Alcuin's solution is a bit more cumbersome than Gauss's, but arguably easier, since multiplication by 100 is easier than multiplication by 101. If, like Alcuin, you're using Roman numerals, do as he did.

The lesson of these two puzzles is this: if you're asked to add up a whole bunch of numbers, don't undertake the challenge literally. Look for the pattern and use it to your advantage.

Overleaf are three splendid counting puzzles for you to put this into practice.

(101)

MIRROR, MIRROR

Which of these two sums is the biggest of them all?

```
  987654321      123456789
  087654321      123456780
  007654321      123456700
  000654321      123456000
  000054321      123450000
  000004321      123400000
  000000321      123000000
  000000021      120000000
+ 000000001    + 100000000
_____    _____


_____    _____
```

(102)

NOUS LIKE GAUSS

The 24 four-digit numbers that include 1, 2, 3 and 4 are as follows, in ascending order. Add them all up please.

1234	1423	2314	3124	3412	4213
1243	1432	2341	3142	3421	4231
1324	2134	2413	3214	4123	4312
1342	2143	2431	3241	4132	4321

(103)

THAT'S SUM TABLE

And now, in two dimensions. You know the drill. What's the sum?

1	2	3	4	5	6	7	8	9	10
2	3	4	5	6	7	8	9	10	11
3	4	5	6	7	8	9	10	11	12
4	5	6	7	8	9	10	11	12	13
5	6	7	8	9	10	11	12	13	14
6	7	8	9	10	11	12	13	14	15
7	8	9	10	11	12	13	14	15	16
8	9	10	11	12	13	14	15	16	17
9	10	11	12	13	14	15	16	17	18
10	11	12	13	14	15	16	17	18	19

The next three puzzles are the maths equivalent of concrete poetry. Each question contains a grid with nine spaces, which should contain the digits from 1 to 9. It is wonderful to see the simplest numerical elements – the non-zero digits – slot together so elegantly.

There are 24,192 possible ways to fit the nine digits into each of these grids. If you tried a new combination every second it would take more than two weeks to try them all. So see if you can find ways to reduce the possible combinations.

(104)

THE SQUARE DIGITS

$$□ - □ = □$$
$$\times$$
$$□ \div □ = □$$
$$=$$
$$□ + □ = □$$

(105)

THE GHOST EQUATIONS

$$□□ \times □ = □□$$
$$□ \times □ = □□$$

(106)

RING MY NUMBER

This puzzle is a three-in-one. Fill in the gaps so that the sum of the digits in each circle is 11. Do it again so they add up to 13, and again so the sum is 14.

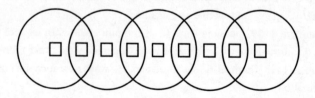

Thomas Dilworth's *The Schoolmaster's Assistant, Being a Compendium of Arithmetic both Practical and Theoretical* was published in 1743, and became an extremely popular maths text book in the UK and the US. It contains the following puzzle.

> 'Says Jack to his brother Harry, "I can place four threes in such manner that they shall make just 34; can you do so too?"'

The answer is $33 + \frac{3}{3} = 34$.

Dilworth's book features the first appearance of puzzles like this one, in which you must arrange four identical digits so that they equal a certain amount. In the three previous questions in this chapter the mathematical operations were given, and the solver had to place numbers between them. Here the numbers are given, and the solver has to place operations between them. The most common variant of this genre is the 'four fours' problem, which was first mentioned a century after Dilworth. As Cupidus Scientiae writes in an 1881 issue of *Knowledge: an Illustrated Magazine of Science*: 'It may be as new to some of the readers ... as it was to myself when first shown the other day that all the numbers to twenty inclusive (and many upwards), with the single exception of nineteen, may be expressed by four fours, using any signs necessary except those of squaring and cubing, in which figures are required.'

The four fours is an unbelievable puzzle: amusing, simple and addictive. It is genuinely surprising how many numbers you can make using no more, and no less, than the digits 4, 4, 4 and 4. However, we need to clarify Cupidus Scientiae's statement about what is possible and which signs are allowed.

THE FOUR FOURS

[1] Use four 4s to make every number from 0 to 9. You are allowed to use only the basic mathematical operations +, −, × and ÷, and brackets. Remember that all four 4s have to be used for each number.

[2] Use four 4s to make every number from 10 to 20. In addition to the basic operations you now may use the √ symbol, the decimal point (so you can write .4), and you can concatenate digits together (so you can write 44, or 444, or even 4.4).

[3] Now you've warmed up, continue from 21 to 50. You are now allowed to use exponentiation (which means that you can write 4⁴), and the factorial symbol ! (so you can write 4!). To get the factorial of a number, we multiply it by every number less than it, so 4! = 4 × 3 × 2 × 1 = 24.

I'll start you off. Here's how you use four 4s to make zero:

$$4 - 4 + 4 - 4 = 0$$

Easy. And here's how you make 1:

$$\frac{(4+4)}{(4+4)} = 1$$

After 50, how much further can we go? A *lot* further. Using only the mathematical operations listed above, we can reach all the numbers to 100, with the exception of 73, 77, 87 and 99, although these are possible with the enterprising use of extra mathematical symbols. For example:

$$(\tfrac{4}{4})\% - \tfrac{4}{4} = 99 \text{ [Because four quarters equal 100 per cent]}$$

In the 1911 edition of *Mathematical Recreations and Essays*, W. W. Rouse Ball writes of the four fours puzzle that he has 'never seen this recreation in print, but it seems to be an old and well-known question.' He says it is possible to reach 170.

By the time of the 1917 edition, however, he had been hard at work. 'If we concede the employment of integral indices and the use of sub-factorials,' he writes, 'we can get to 877,' adding that 'with four "1s" we can get to 34, with four "2s" to 36, with four "3s" to 46, with four "5s" to 36, with four "6s" to 30, with four "7s" to 25, with four "8s" to 36, and with four "9s" to 130.' How appropriate that four *fours* gets us the f(o)urthest.

Has anyone gone any higher? Yes! In the following decade Paul Dirac, the mathematical physicist we met at the end of the last chapter, solved the four fours for all numbers up to infinity.

Dirac's solution, in fact, was for the 'four twos' problem, which was doing the rounds at Cambridge at the time, but his solution also works for the four fours. If logarithms are allowed, any number n can be expressed by: $\log_{\sqrt{4}/4} (\log_4 \sqrt{\ldots} \sqrt{4})$ where n is the number of square root signs in the above term. (Don't worry if you don't understand logarithms; you only need to appreciate the stunning brevity and easy scalability of the answer.) Dirac loved mathematical puzzles and must have been thrilled to generalise such a well-known problem with such an ingenious formula. 'He had rendered the game pointless,' writes Graham Farmelo in his Dirac biography *The Strangest Man.*

In 1882, a year after *Knowledge* magazine first referred to the four fours puzzle, the US puzzle impresario Sam Loyd published Our Columbus Problem, the most wicked and absurd of the 'have-numbers-find-operations' genre. He offered a prize of $1,000 for the best answer – about £20,000 in today's money – and received only two correct answers out of several million responses. Well, that's what he said anyway. Loyd was as talented at self-promotion as he was at inventing puzzles. I'm including the problem here for historical completeness, not because I think you will be able to do it. Go on, prove me wrong!

(108)

OUR COLUMBUS PROBLEM

Use the following seven digits and eight dots
$\cdot 4 \cdot 5 \cdot 6 \cdot 7 \cdot 8 \cdot 9 \cdot 0 \cdot$
to make an addition with an answer that is as close to 82 as possible.

The dots can be used in two ways: [1] as decimal points; and [2] as the symbols that represent repeating decimals, which are placed on top of a digit or digits. If a single digit has a dot on it, the digit is repeated endlessly, so $\frac{1}{3}$ can be written $.\dot{3}$ rather than $.3333....$ If two digits have dots on them, the sequence beginning with the first digit and ending with the second digit is repeated endlessly. So, $\frac{1}{7}$ can be written $.\dot{1}4285\dot{7}$, rather than $.142857142857142857...$

That's the starters over with. But before we move on to other courses, here's a palate cleanser.

(109)

THREES AND EIGHTS

Can you make 24 using 3, 3, 8 and 8?
   You are allowed to use only the basic mathematical operations $+$, $-$, $\times$ and $\div$, and brackets.

The following number puzzle went viral a few years ago. It was accompanied by the words: 'This problem can be solved by pre-school children in five to ten minutes, by programmers in an hour and by people with higher education ... well, check it yourself!' I'm not sure this statement was scientifically verified, but it certainly made you want to solve the problem.

(110)

CHILD'S PLAY

8809 = 6	5555 = 0
7111 = 0	8193 = 3
2172 = 0	8096 = 5
6666 = 4	1012 = 1
1111 = 0	7777 = 0
3213 = 0	9999 = 4
7662 = 2	7756 = 1
9313 = 1	6855 = 3
0000 = 4	9881 = 5
2222 = 0	5531 = 0
3333 = 0	2581 = ?

Numbers can describe quantities.

One paragraph. Six words. Three sentences.

But when numbers are in a list they can also express an *order.*

The next three problems involve sequences of numbers. In each the puzzle is to work out the rule and discover what comes next.

(111)

FOLLOW THE ARROW 1

77 → 49 → 36 → 18 → ?

The following puzzle is by Nob Yoshigahara, who devised the problem that opens this book. It is magical to find a sequence that doubles back on itself like this.

FOLLOW THE ARROW 2

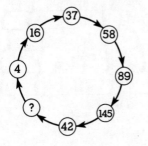

FOLLOW THE ARROW 3

$10 \rightarrow 9 \rightarrow 60 \rightarrow 90 \rightarrow 70 \rightarrow 66 \rightarrow$?

I write about maths, so I like numbers and I like words. Predictably, I'm very fond of puzzles that link numbers and words.

The next one is a brilliantly simple idea: imagine if numbers were listed in alphabetical order.

(114)

DICTIONARY CORNER

A dictionary lists the whole numbers from one to one quadrillion (that's 1 to 1,000,000,000,000,000) in alphabetical order. Find:

The first entry.
The last entry.
The first odd entry.
The last odd entry.

For clarification, this dictionary has the following rules:

[1] Words are written the American way, which omits the 'and'. For example, 2001 is listed as 'two thousand one', not 'two thousand *and* one'.

[2] The alphabetic representation of 100 is '*one* hundred', for 1000 is '*one* thousand', and so on for the bigger numbers.

[3] Spaces and hyphens are ignored; so, for example, 'fourteen' precedes 'four trillion'.

Sam Loyd was one of the first puzzle inventors to devise problems in which words represent numbers. His General Store Puzzle lists the following items:

```
        C H E S S
        C A S H
      B O W W O W
        C H O P S
    A L S O P'S
  P A L E A L E
        C O O L
        B A S S
        H O P S
        A L E S
        H O E S
    A P P L E S
        C O W S
    C H E E S E
    C.H.S O A P
      S H E E P
```
```
A L L W O O L
```

Use the ten letters of PEACH BLOWS (a variety of potato) as a key that refers to the ten digits 1, 2, 3, 4, 5, 6, 7, 8, 9, and 0 in order, i.e., P = 1, E = 2, A = 3, C = 4, H = 5, and so on. The word CHESS, therefore, becomes the number 45,200, and CASH becomes 4,305. If we think of the list of the 16 words above as an addition of the numbers they represent, then the correct answer to the sum is the final word, ALLWOOL, or 3,779,887.

Loyd's puzzle is clever but a bit woolly: too many numbers to make it fun. The format, however, in which meaningful phrases have a unique solution when the letters are substituted for digits – and which is now known as alphametics, cryptarithms or verbal arithmetic – was perfected by Henry Ernest Dudeney. In 1924 he published the following puzzle, still one of the best of the genre:

$$
\begin{array}{r}
S\ E\ N\ D \\
+\ \ M\ O\ R\ E \\
\hline
M\ O\ N\ E\ Y
\end{array}
$$

To solve it, you need to find the digits for which the sum is correct (with the provisos that each letter represents a unique digit, and the leftmost letters are not 0).

Loyd was sixteen years older than Dudeney, and the only one of Dudeney's puzzlist contemporaries comparable in output and originality. The men corresponded across the Atlantic, although Dudeney broke off the friendship when he discovered that Loyd was passing off his puzzles under his own name. In fact, the personalities of both men reflected national stereotypes: Loyd, an endlessly energetic puzzle-producing machine, was an entrepreneur, patenting his best ideas, offering prize money to solvers and embellishing his biography as his reputation grew. Dudeney was a grumpy, pipe-smoking, provincial Brit.

SEND MORE MONEY is so well known I'll present it here as a bonus puzzle. Here's how you start. The letter M must be 1, since when two four-digit numbers are added together to make a five-digit number, that five-digit number can only begin with a 1. (Consider the largest four-digit number, 9,999. When added to itself the answer is 19,998, which begins with a 1. So it is not possible for two four-digit numbers to sum to a five-digit number beginning with a 2 or above.)

$$
\begin{array}{r}
\text{S E N D} \\
+ \quad \text{1 O R E} \\
\hline
\text{1 O N E Y}
\end{array}
$$

In order for S + 1 = 1O (where O is a capital o, not a zero), either S = 9, or S = 8 and there is a 1 carried from the hundreds column. Let's imagine S = 8 and there is a carry, which means the capital o would be a 0. The sum becomes:

$$
\begin{array}{r}
\overset{1}{8} \text{E N D} \\
+ \quad \text{1 O R E} \\
\hline
\text{1 O N E Y}
\end{array}
$$

I've put the carry above the 8 to make it clearer. If this sum is correct, then the hundreds column tells us either that E + O = 10 + N, or there is a carry from the tens column and 1 + E + O = 10 + N. (The 10 in these equations represents the carry to the thousands column.) In the former case this means that the difference between E and N is 10, which is impossible since E and N are both less than 10. The latter case is also impossible since it would mean that E − N = 9. The only two values this works for is E = 9 and N = 0. But capital O is already 0, and two different letters cannot represent the same digit. So, S = 9, and I'll leave the rest to you. (The answer is in the back.)

There are many brilliant alphametic puzzles, but the following is a particular favourite of mine, since it is almost a perfect quote from *Macbeth*, missing only the 'and' that separates 'toil' and 'trouble'. However, if you place the addition sign cleverly...

(115)

THE THREE WITCHES

Find the digits for which the following sum is correct:

```
D O U B L E
D O U B L E
  T O I L +
─────────────
T R O U B L E
```

Here's another alphametic puzzle with an irresistible twist.

ODDS AND EVENS

In the long multiplication shown below, each E is an even digit and each O is an odd digit. In other words, each E is either 0, 2, 4, 6 or 8, and each O is either 1, 3, 5, 7 or 9. Just because two digits are Es does not mean they are necessarily the same digit, although in some cases they might be. Can you reconstruct the multiplication? (The space in the units column represents a zero in that position. The symbol is not included here since it would be too confusing with all the Os. Also, this position in a long multiplication is always a zero so there is nothing to deduce.)

```
      E E O
    ×   O O
    ─────────
    E O E O
    E O O
    ─────────
    O O O O O
```

The previous problem, devised in the early 1960s by William Fitch Cheney, a mathematics professor and magician, first appeared in Martin Gardner's *Mathematical Games* column in *Scientific American*. If Sam Loyd was America's greatest puzzle inventor, Gardner was America's greatest popula- riser of mathematical puzzles. Through his *Scientific American* column, which ran for more than twenty years, and his dozens of books, Gardner assembled and exposited an unrivalled library of mathematical puzzles. He also became the hub of an extensive, informal network of puzzle enthusiasts – such as Fitch Cheney – whose best ideas often debuted in his columns.

The next puzzle is by Lee Sallows, a master of mathematical wordplay whose work was also given a wide audience by Martin Gardner. For me, this fantastically ingenious self-enumerating crossword is a work of art.

(117)

THE CROSSWORD THAT COUNTS ITSELF

Each of the entries in the crossword shown below is of the form:

   [NUMBER][SPACE][LETTER][S]

and accurately describes the number of times that particular letter appears in the grid.

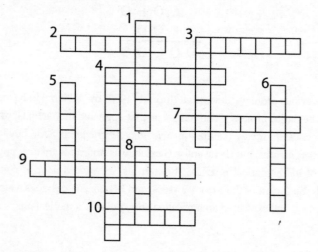

For example, if there was a single 'Q' in the grid, one of the entries would be:

'ONE Q'

If there were five 'P's and seventeen 'E's in the grid, two of the entries would be:

'FIVE PS'

'SEVENTEEN ES'

In other words, every entry is a number word followed by a blank space followed by the letter in question, and followed by an 's' if it is plural. And each entry states the correct number of times that letter appears in the grid.

Fill in the crossword using logic alone.

This puzzle is beautifully self-contained: only 12 letters are used in the entire grid, and each one has its own entry.

To get you started, here's how to fill in the first three letters. The entry 8 Down has only five spaces, so it must be of the form ONE *, where * is a single letter. (Remember that all entries with numbers bigger than one will need at least six spaces because of the added 'S' for the plural.)

Now you're on your own.

If a crossword can count itself, how might a number count itself?

Here's one way. We say that the number 1,210 counts itself because its first digit (which is 1) describes how many 0s it has, the second digit (2) describes how many 1s it has, the third digit (1) describes how many 2s it has, and the fourth digit (0) how many 3s. The self-descriptive properties of 1,210 are especially clear when the number is written in a grid:

0	1	2	3
1	**2**	**1**	**0**

Each digit in the second row describes how many times the number above it appears in the second row.

Numbers like 1,210, whose left-most digit describes how many 0s it has, whose leftmost-but-one digit describes how many 1s it has, whose leftmost-but-two digit describes how many 2s it has, and so on, are called 'autobiographical numbers'. There are only two four-digit autobiographical numbers: 1,210 and 2,020.

The only five-digit autobiographical number is 21,200:

0	1	2	3	4
2	**1**	**2**	**0**	**0**

It has two 0s, one 1, two 2s, no 3s and no 4s.

Got it? Now we're ready.

(118)

AN AUTOBIOGRAPHY IN TEN DIGITS

Find the only ten-digit autobiographical number.

The number will fit in the second row of the grid below. Each of its digits will describe the number of times the digit above it appears in the second row.

0	1	2	3	4	5	6	7	8	9

A number that includes all ten digits 1, 2, 3, 4, 5, 6, 7, 8, 9 and 0 is called a *pandigital* number, such as 1,234,567,890. (The leftmost digit of a pandigital number must be non-zero.)

(119)

PANDIGITAL PANDEMONIUM

How many ten-digit pandigital numbers are there?

Here's an interesting fact about ten-digit pandigital numbers: they are all divisible by 3.

We can show this using the 'divisibility by 3' test, which you may remember from school. If we add the digits of a particular number and get an answer that's divisible by 3, then the original number is divisible by 3.

A ten-digit pandigital number must include every digit only once. So if we add up its digits, we get $1 + 2 + 3 + 4 + 5 + 6 + 7 + 8 + 9 + 0 = 45$, which is divisible by 3. So all pandigitals are also divisible by 3. Lovely.

Less well-known divisibility tests are:

Divisibility by 4. If the last two digits of a number are divisible by 4, then the entire number is divisible by 4.

Divisibility by 8. If the last three digits of a number are divisible by 8, then the entire number is divisible by 8.

You might want to think about why these two tests are true.

Or not. Either way, they will come in handy in the next puzzle.

PANDIGITAL AND PANDIVISIBLE

Find the ten-digit pandigital number *abcdefghij* where:

a is divisible by 1
ab is divisible by 2
abc is divisible by 3
abcd is divisible by 4
abcde is divisible by 5
abcdef is divisible by 6
abcdefg is divisible by 7
abcdefgh is divisible by 8
abcdefghi is divisible by 9
abcdefghij is divisible by 10

This is a fantastically elegant puzzle because the conditions provided in the question give a unique solution. You will need a calculator, but that doesn't spoil the journey.

Think of a number.

The number must be a three-digit number in which the first and last digits differ by at least two. So, for example, 258.

Reverse this number and calculate the difference between the two numbers you have:

In my example, 852 − 258 = 594.

Add this number to its reverse: 594 + 495.

The answer is 1,089.

Now try this process again with a different number: reverse it, calculate the difference, add this number to its reverse, and…

You guessed it! The answer is 1,089.

Whichever three-digit number you start with you will always get 1,089, which is pretty cool the first time you try it out.

And there is another reason why 1,089 is arithmetically notable.

(121)

1089 AND ALL THAT

When you multiply 1089 by 9 the number reverses itself:
$$1089 \times 9 = 9801$$
Can you find the four-digit number that reverses itself when it's multiplied by 4? In other words, the number *abcd* such that:
$$abcd \times 4 = dcba$$

The number 102,564 also changes in a pleasing way on multiplication by 4:

$$102564 \times 4 = 410256$$

Did you spot it? The final digit of 10256**4** becomes the first digit of **4**10256, while all the other numbers stay the same. In other words, when you multiply 102546 by 4, the answer is the same set of digits, but with the rightmost digit of the original number now in the leftmost position of the answer.

The same transformation happens in the following equation.

$$142857 \times 5 = \mathbf{7}14285$$

The rightmost digit of the first number, in this case 7, becomes the first digit of the answer, and all the other digits remain the same.

(122)

BACK TO FRONT

Can you find a number *N* such that when it is multiplied by 2 the answer has exactly the same digits as *N*, in the same order, except that the rightmost digit of *N* is now the first digit of the answer?

(In other words, what happens to *N* when it is doubled is the same as what happens to 102,564 when it is multiplied by 4 and 142,857 when it is multiplied by 5.)

Freeman Dyson, the eminent British physicist, heard this question being discussed in the cafeteria of a science conference.

'Oh, that's not difficult,' he chipped in, 'but of course the smallest such number is 18 digits long.'

According to the *New York Times* his colleagues were astonished. No one had a clue how 'Freeman could have known such a fact or, even more terrifying, could have derived it in his head in about two seconds.'

Yet Dyson was correct, and the answer requires mathematics a primary school pupil would understand.

We're approaching the end of the book, and the numbers are getting bigger. So big, in fact, that there is no space for them.

(123)

THE NINTH POWER

The following nine numbers are the last four digits of 31^9, 32^9, 33^9, 34^9, 35^9, 36^9, 37^9, 38^9 and 39^9, but listed in a random order.

Can you list them in ascending order?

... 2848

... 5077

... 1953

... 6464

... 8759

... 8832

... 0671

... 1875

... 8416

My next number looks down its nose at 39^9 for not trying hard enough.

(124)

WHEN I'M SIXTY-FOUR

Estimate 2^{64}.

Our final number considers even 2^{64} an insignificance.

A LOT OF NOTHING

How many zeros does the number 100! have at the end of it?

The number 100! means you multiply 100 by every whole number less than it, as I discussed on page 172. So it's equal to $100 \times 99 \times 98 \times 97 \times 96 \times \ldots \times 3 \times 2 \times 1$. You are not expected to *actually* calculate the answer (it has 158 digits). Instead, use mathematical insight to work out what it means to have zeros at the end.

ANSWERS

Ten tasty teasers

Are you smarter than an 11-year-old?

1) D

The three images show that there are six letters on the cube: I, K, M, O, U and P. Since a cube has six faces, these must be the only letters on the cube. The first view of the cube shows that I and M each share a side with K. The second view shows that O and U each share a side with K. Only four faces can share a side with K. If the cube is positioned such that K is on the top, as in the first image, then I is adjacent in a clockwise direction from M. We can deduce from the second image that when K is on the top the O is adjacent in a clockwise direction from U, so the clockwise order of faces around K must be M-I-U-O. Thus M is opposite U.

2) D

After nine lies, the length of his nose will be $2^9 \times 5\text{cm} = 512 \times 5\text{cm} = 25.6\text{m}$. This is close to the length of a tennis court, which is 23.8m. This distance is well short of the physical maximum for Pinocchio's nose, however, according to a 2014 report from the University of Leicester's Centre for Interdisciplinary Science. According to its calculations, if the puppet's wooden head is 4.18kg, with a 6g nose of initial length one inch (2.54cm), the nose will break only after 13 fibs, when it reaches a length of 208m.

3) C

Eighteen has eight letters and is not a multiple of 8.

4) D

Well, for starters Amy is to the left of both Ben and Chris. So the three are in the order Amy, Ben, Chris, or Amy, Chris, Ben. That's all we know, so D is certainly true. None of the other statements *must* be true, even though B *may* be true.

5) E

You could find this out by trial and error. Or maybe you worked out the rule: for it to be possible to draw a figure without taking the pen off the paper and without drawing along an existing line, there must be at most two points in the figure at which an odd number of lines meet. Only E satisfies this condition.

6) B

I hope at least that you know your seven times table! If you do it will come as no surprise that 7 divides 35, so 7 divides 350,000. Seven also divides 49, so 7 divides 4,900. Since 354,972 = 350,000 + 4,900 + 72, we are left to find the remainder of 72 divided by 7. Since $7 \times 10 = 70$, the remainder is 2.

7) C

There must be at least two boys, because if there was only one boy he would have no brother, which contradicts the question. Likewise, there must be at least two girls, so the smallest number is four.

8) E

Just do this fun multiplication on the back of an envelope and be done with it.

$$
\begin{array}{r}
9\ 8\ 7\ 6\ 5\ 4\ 3\ 2\ 1 \\
\times\ 9 \\
\hline
8\ 8\ 8\ 8\ 8\ 8\ 8\ 8\ 9
\end{array}
$$

9) A

I hope there was some space left on that envelope. The calculations required are: $p = 105 - 47 = 58$; $q = p - 31 = 58 - 31 = 27$; $r = 47 - q = 47 - 27 = 20$; $s = r - 13 = 20 - 13 = 7$; $t = 13 - 9 = 4$; $x = s - t = 7 - 4 = 3$.

10) A

It would have been churlish of me not to have included a long division. $20/11 = 1.818181...$, so only two different digits appear.

Cabbages, Cheating Husbands and a Zebra

LOGIC PROBLEMS

② THREE FRIENDS AND THEIR SISTERS

The nine-crossing solution is as follows. The macho set-up of the problem is mitigated by the fact that women are rowing on at least six crossings – and possibly on all of them. The outline of the strategy is to take the first pair, then the second, and then the third, with the brothers always landing before the sisters.

LEFT BANK		RIGHT BANK
1) B_2 B_3 S_2 S_3	$B_1 S_1$ →	
2) B_2 B_3 S_2 S_3	← S_1	B_1
3) B_2 B_3 S_3	$S_1 S_2$ →	B_1
4) B_2 B_3 S_3	← S_2	B_1 S_1
5) B_3 S_3	$B_2 S_2$ →	B_1 S_1
6) B_3 S_3	← S_2	B_1 S_1 B_2
7) B_3	$S_2 S_3$ →	B_1 S_1 B_2
8) B_3	← S_3	B_1 S_1 B_2 S_2
9)	$B_3 S_3$ →	B_1 S_1 B_2 S_2

The second step is not allowed with the stricter conditions, since when the sister from the first pair returns to the left bank she will be unaccompanied and in the presence of unrelated men. In this case, the quickest solution is eleven crossings. The lesson of the wolf, goat and cabbages puzzle was that to get everything across we have to take one item across the river, back and across again. Here we have to ferry across every sister, ferry them all back, and ferry them all across again.

Here's one way to do it:

LEFT BANK		RIGHT BANK
1) B_1 B_2 B_3 S_3	S_1 S_2 →	
2) B_1 B_2 B_3 S_3	← S_2	S_1
3) B_1 B_2 B_3	S_2 S_3 →	S_1
4) B_1 B_2 B_3	← S_2	S_1 S_3
5) B_2 S_2	B_1 B_3 →	S_1 S_3
6) B_2 S_2	← B_3 S_3	B_1 S_1
7) S_2 S_3	B_2 B_3 →	B_1 S_1
8) S_2 S_3	← S_1	B_1 B_2 B_3
9) S_3	S_1 S_2 →	B_1 B_2 B_3
10) S_3	← B_3	B_1 B_2 S_1 S_2
11)	B_3 S_3 →	B_1 B_2 S_1 S_2

This solution is the one that Alcuin gave, and (in the version of the puzzle where the couples are husband and wife) which is encoded in the Latin hexameter. A rough translation:

> *Women, woman, women, wife, men, man and wife,*
> *Men, woman, women, man, man and wife.*

③ CROSSING THE BRIDGE (WITH A LITTLE HELP FROM MY FRIENDS)

The strategy I mentioned in the main text is for John, the fastest walker, to take each companion across one by one. He takes Paul across in 2 minutes, and returns in 1. He takes George across in 5 and returns in 1. Finally, he takes Ringo across in 10. The total time is 2 + 1 + 5 + 1 + 10 = 19 minutes.

At first it seems self-evident that this strategy is the best. Why not use the fastest man at all times? Well, because it makes sense to bundle the slowcoaches together. Here's how:

[1] John takes Paul across in 2 minutes and returns in 1. Same as before.

[2] George and Ringo now cross together, which takes them 10 minutes.

[3] They hand the torch to Paul, who returns across the bridge, adding an extra 2 minutes.

[4] John and Paul make the final crossing together, in another 2 minutes.

The total is 2 + 1 + 10 + 2 + 2 = 17 minutes.

This puzzle is fantastic because what seems like the wrong thing to do, which is to reduce John's participation, is actually the right thing to do. The solution provides a real 'wow!'.

To get more of a feel for why the best solution is to make the two slowest go together, imagine if the puzzle had instead stated that John takes 1 minute and Paul takes 2 minutes, but that George takes 24 hours and Ringo 24 hours and 1 minute. Now it seems much more obvious for George and Ringo to share the torch, since that requires only one 24-hour crossing.

④ THE DOUBLE DATE

The two sons are both uncle and nephew to each other. For a 'simple' puzzle it is surprisingly brain-muddling! Let's call the two men Albert and Bernard, and their sons Steve and Trevor. I've drawn the family tree opposite.

Bernard and Steve are half-brothers because they have the same mother. Bernard's son Trevor is therefore Steve's nephew.

Likewise, Albert and Trevor are half-brothers, so Steve is Trevor's nephew.

The confusing family ties get even more twisted when we consider that Bernard's mum is married to Albert, so Bernard's mum's stepmother is Albert's mother. This makes Albert's mother Bernard's step-grandmother. Bernard is therefore married to his step-grandmother, and is therefore step-grandfather to himself.

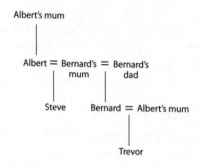

⑤ THE DINNER PARTY

There is only one guest.

The diagram below reveals the relations in this strange family. The Governor's father is Mr C, so the guest is the Governor's father's brother-in-law. Likewise, each of the other descriptions of the guest represents a different path from the Governor to him: via the Governor's brother (Mr E), via his father-in-law (Mr B), and via his brother-in-law (Mr D).

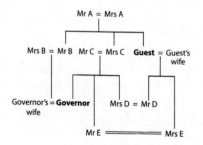

⑥ LIARS, LIARS

We are looking for the combination of truth-telling and lying that does not lead to a contradiction.

Let's say Berta is telling the truth. It follows that Greta is lying. So Rosa must be telling the truth. But if she is telling the truth, then *both* Berta and Greta are lying, which is a contradiction. So Berta is not telling the truth.

Now if Berta is lying, then Greta is telling the truth, which means Rosa is a liar. If she is a liar then at least one of Berta and Greta is telling the truth, which is a valid statement. So the combination of Berta and Rosa being liars and Greta being the one who tells the truth is logically consistent, and is the answer.

⑦ SMITH, JONES AND ROBINSON

I told you that the guard's nearest neighbour earns exactly three times as much as the guard does, which means that the guard's nearest neighbour cannot be Mr Jones, because his salary is indivisible by three. But neither can the guard's nearest neighbour be Mr Robinson, because the guard lives between Leeds and Sheffield and Mr Robinson is in Leeds. So the guard's nearest neighbour, and fellow resident of 'between Leeds and Sheffield', must be Mr Smith. We can put a tick in the top right of the right-hand box, as shown below, and deduce therefore that Mr Jones is in Sheffield because that is the only option remaining.

The guard's namesake lives in Sheffield. We know that Mr Jones lives in Sheffield. So the guard must be Jones. We can tick Jones/ guard, as shown below left, and cross off the other boxes in the same row and column since Jones has no other occupation and the others are not the guard.

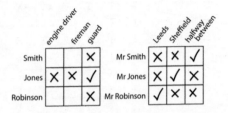

The clue that Smith beats the fireman at billiards reveals that Smith *is not* the fireman. (Robinson must be the fireman.) So we can cross off Smith/fireman. We already know that Smith is not the guard. So by elimination Smith is the engine driver.

(8) ST DUNDERHEAD'S

We find out who went to the cinema by going through the group one by one, each time assuming that the girl in question was at the cinema, and counting the number of girls who are telling untruths.

For example, let's assume that Joan Juggins went to the cinema. Her statement that it was Joan Twigg is untrue, as is Gertie Gass's. However, Bessie and Sally must be telling the truth. When we mark this up in a table the pattern is easier to see. In the table below, the first row shows the truth/falsity of statements when Joan Juggins is in the cinema, the second when Gertie Gass is in the cinema, and so on. The final column is the total number of false, or untrue, statements. So if T is true, and F false, the first row begins F, F, T, T, and once completed looks like this:

The statements

	JJ	GG	BB	SS	MS	DS	KS	JT	JF	LL	FF	#F
JJ	F	F	T	T	F	T	T	F	T	F	T	5
GG	F	T	F	F	F	T	T	F	T	F	T	6
BB	F	F	T	T	T	F	T	T	F	F	T	5
SS	F	F	T	T	F	T	T	T	F	F	T	5
MS	F	F	T	T	F	T	F	F	T	T	F	6
DS	F	F	T	T	F	F	F	F	T	F	F	8
KS	F	F	T	T	F	T	F	F	T	T	F	6
JT	T	F	T	F	F	T	T	F	F	F	T	6
JF	F	F	T	T	F	T	T	F	T	F	T	5
LL	F	F	T	T	F	T	T	F	T	F	T	5
FF	F	F	T	T	F	T	T	F	T	F	T	5

In the cinema

If at least seven statements are untrue, then Dorothy Smith must be the secret cinephile.

(9) A CASE OF KINSHIP

We have five men: Jenkins, Tomkins, Perkins, Watkins and Simkins. For simplicity's sake, let's call them J, T, P, W and S. And we have five women, who are both wives and mothers to specific men (although a woman is never a wife and a mother of the same man; love in Kinsleydale is weird, but not that weird). Let's identify the women by their blood relationships to the others, and let's use lower-case letters, so j is J's mother, t is T's mother, and so on.

We're going to draw a grid. The top line contains the men and the bottom line their wives, so to begin with it's blank. If Jenkins's stepson is Tomkins, this means that Mrs Jenkins is Tomkins's mum, so we can put t under J.

Man	J	T	P	W	S
Wife	t				

We also know that Tomkins is the stepfather of Perkins, which means that Mrs Tomkins is Perkins's mum. Under T we can put p.

Man	J	T	P	W	S
Wife	t	p			

The set-up tells us that Jenkins's mother is a friend of Mrs Watkins. So we know that Mrs Watkins is *not* Jenkins's mum. Since Mrs Watkins cannot be Watkins's mum, by a process of elimination she must be Simkins's mum.

Man	J	T	P	W	S
Wife	t	p		s	

Finally, we are told that Mrs Watkins's husband's mother (i.e., Watkins's mother) is a cousin of Mrs Perkins. So Perkins's wife is not Watkins's mother. If Perkins's wife is not Watkins's mother, then she can only be Jenkins's mother. Again by a process of elimination, Simkins's wife must be Watkins's mum.

Man	J	T	P	W	S
Wife	*t*	*p*	*j*	*s*	*w*

So Simkins's stepson is Watkins.

⑩ THE ZEBRA PUZZLE

It's a grid puzzle, so draw a grid. There are five houses and five attributes, so the grid will look as it does below.

We'll fill in the blanks statement by statement.

Statement **9** says that milk is drunk in the middle house, so we can put *milk* in column three; **10** that the Dane is in the first house, so we can put *Dane* in column one; and **15** that the house next to the Dane's is blue, so we can put *blue* in column two.

	HOUSE 1	HOUSE 2	HOUSE 3	HOUSE 4	HOUSE 5
COLOUR		Blue			
NATIONALITY	Dane				
PET					
BEVERAGE			Milk		
SHOE					

Statement **6** tells us that the green and ivory houses are next to each other. So the first house cannot be green or ivory. But neither can the first house be red, since according to **2** the Scot is in the red house, and we know the Dane is in the first house. We can infer that the first house is yellow. According to **8**, brothel creepers are worn there, and from **12** we know that the second house has a horse.

	HOUSE 1	HOUSE 2	HOUSE 3	HOUSE 4	HOUSE 5
COLOUR	Yellow	Blue			
NATIONALITY	Dane				
PET		Horse			
BEVERAGE			Milk		
SHOE	B. creepers				

What does the Dane drink? Not coffee because of **4**, not tea because of **5**, not milk because of **9**, and not orange juice because of **13**. So the Dane must drink water.

And who lives in the second house? It's not the Scot, since the house is blue, and it's not the Greek, since the pet is a horse. It is either the Bolivian or the Japanese. But if it is the Japanese, what does he drink? Not water, not milk, not coffee (by **4**) and not tea (by **5**). So the Japanese must drink orange juice. But then according to statement **13** he wears slippers, contradicting **14**, which says that he wears Havaianas. So the Bolivian must be in the second house, where he drinks tea.

	HOUSE 1	HOUSE 2	HOUSE 3	HOUSE 4	HOUSE 5
COLOUR	Yellow	Blue			
NATIONALITY	Dane	Bolivian			
PET		Horse			
BEVERAGE	**WATER**	Tea	Milk		
SHOE	B. creepers				

According to statement **6** the green and ivory houses are neighbours, which means that the red house is either the third or the fifth house. Imagine it is the fifth house. Then the Scot lives there, and by **4** he drinks orange juice, and by **13** he wears slippers. But if this is the case, who wears brogues and owns snails, as stated in **7**? Not the Dane, who wears brothel creepers, nor the Bolivian, who has a horse, nor the Greek, who by **3** has a dog, nor the Japanese, who according to **14** wears Havaianas. No one can! So we can conclude that the third house is the Scot's red house, and therefore the fourth and fifth, by **6**, are ivory and green. According to **4** coffee is drunk in the fifth house, so orange juice must be drunk in the fourth. And according to **13**, slippers are also in the fourth.

	HOUSE 1	HOUSE 2	HOUSE 3	HOUSE 4	HOUSE 5
COLOUR	Yellow	Blue	Red	Ivory	Green
NATIONALITY	Dane	Bolivian	Scot		
PET		Horse			
BEVERAGE	**WATER**	Tea	Milk	Orange juice	Coffee
SHOE	B. creepers			Slippers	

The Japanese must be in the fifth house wearing Havaianas, since by **14** he cannot be in the fourth one, where the Greek must live with his dog.

	HOUSE 1	HOUSE 2	HOUSE 3	HOUSE 4	HOUSE 5
COLOUR	Yellow	Blue	Red	Ivory	Green
NATIONALITY	Dane	Bolivian	Scot	Greek	Japanese
PET		Horse		Dog	
BEVERAGE	**WATER**	Tea	Milk	Orange juice	Coffee
SHOE	B. creepers			Slippers	Havaianas

The rest of the table now fills itself: the snail-owning brogue wearer must be the Scot. So the Bolivian wears Birkenstocks, and by **11** the Dane must have the fox. The last remaining square must be the zebra, the owner of which is Japanese.

	HOUSE 1	HOUSE 2	HOUSE 3	HOUSE 4	HOUSE 5
COLOUR	Yellow	Blue	Red	Ivory	Green
NATIONALITY	Dane	Bolivian	Scot	Greek	Japanese
PET	Fox	Horse	Snails	Dog	**ZEBRA**
BEVERAGE	**WATER**	Tea	Milk	Orange juice	Coffee
SHOE	B. creepers	Birkenstocks	Brogues	Slippers	Havaianas

There are other ways you could have filled the table in – but the final version must always look like this one!

⑪ CALIBAN'S WILL

How do we even start? Let's restate the three statements:

[1] No person who has seen Caliban in a green tie is to choose before Low.

[2] If Y.Y. was not in Oxford in 1920 the first chooser never lent Caliban an umbrella.

[3] If Y.Y. or 'Critic' has second choice, 'Critic' comes before the one who first fell in love.

We are trying to find the order in which Low, Y.Y. and 'Critic' chose Caliban's books. The key here is that every statement is necessary for the solution. In other words, every statement must contain useful information. If at least one of the statements provides us with no information in reaching a solution, then that solution is wrong.

In order for [1] to provide us with information, at least one out of Y.Y. and 'Critic' has seen Caliban in a green tie. If no one has seen him in a green tie, then the statement is redundant. We can deduce therefore that Low cannot be the third-chooser, as he must be followed by those who saw Caliban in a green tie.

Now to statement [2]. If Y.Y. was in Oxford then [2] provides no information about the order, so we can say that Y.Y. was not in Oxford. And if no one lent Caliban an umbrella the statement is superfluous. So someone lent Caliban an umbrella.

So who lent Caliban an umbrella? If Low lent Caliban an umbrella, then from [2] Low is not first. We know Low is not last from [1], so this would put Low second. But if Low is second then [3] tells us nothing new and the puzzle is not solvable. So Low did not lend Caliban an umbrella

If both Y.Y. and 'Critic' lent Caliban an umbrella, then from [2] Low is first, and from [3] 'Critic' is second and Y.Y. is third, so [1] is superfluous. So either Y.Y. or 'Critic' lent Caliban an umbrella, but not both. Likewise, if both Y.Y. and 'Critic' saw Caliban in a green tie, from [1] Low is first and [2] is superfluous. So either Y.Y. or 'Critic' saw Caliban in a green tie, but not both.

Let's say that Y.Y. saw Caliban in a green tie *and* lent him an umbrella. From [1] we know that Y.Y. cannot be first, which if true makes [2] superfluous. So if Y.Y. saw Caliban in a green tie, then he cannot have lent him an umbrella, which means that 'Critic' lent him an umbrella. Likewise, if 'Critic' saw Caliban in a green tie, the same logic applies and Y.Y. must have lent Caliban an umbrella.

In both of these cases, Low must be first. And, if this is the case, from [3] Y.Y. must be the one who first fell in love, so the final order is Low, 'Critic', Y.Y.

Caliban's Will: update

In the wake of the hardback edition some readers have argued that the puzzle is solvable using only [1] and [3]. Their argument is as follows: [1] shows Low is first or second. In order for [3] to not be superfluous, then either Y.Y. or 'Critic' is second, which means that the order must be Low, 'Critic', Y.Y. This reasoning is fallacious. Here's why. Let's assume that Y.Y. saw Caliban in a green tie, 'Critic' lent Caliban an umbrella and Y.Y was the one who first fell in love. From [1] we are left with the following possible orders: (i) 'Critic', Low, Y.Y, (ii) Low, Y.Y., 'Critic', (iii) Low, 'Critic', Y.Y. From [3] we can eliminate (ii), leaving us with two options. The puzzle as it stands is not yet solved, but we have already used [3], so it is evidently not superfluous. In other words, we cannot logically deduce that either Y.Y. or 'Critic' is second from [1] and [3] alone, since [3] has meaning even when neither Y.Y nor 'Critic' are second. As a result, we must obtain information in [2] to solve the problem.

⑫ TRIANGULAR GUNFIGHT

The truel is a logical jewel. It produces the brilliantly counter-intuitive – and pacifist – result that Ugly's best chance of survival is to kill no one first.

Ugly must certainly not aim at Bad. If he kills him, Good will kill Ugly with a certainty of 100 per cent. Thud.

What if Ugly aims at Good, to eliminate the sharpest shot straight away? If Ugly kills Good, this leaves Ugly and Bad to shoot it out between themselves. In this scenario, while there is no certainty that Ugly will be killed, the odds are badly stacked against him. Bad is a better marksman *and* he gets to shoot first. Ugly's odds of survival, in fact, are $\frac{1}{7}$, or 14 per cent.

(This is worked out by calculating that Bad's chance of winning in one shot is $\frac{2}{3}$, in two shots $(\frac{2}{3})(\frac{1}{3})(\frac{2}{3})$, in three shots $(\frac{2}{3})(\frac{1}{3})(\frac{2}{3})(\frac{1}{3})(\frac{2}{3})$, and so on. Add up this infinite series and you get $\frac{6}{7}$. The chance of Ugly surviving is therefore $\frac{1}{7}$.)

If Ugly *misses* Good, then it's Bad's turn to shoot and he will aim at Good, with a ⅔ chance of killing him. If he succeeds, we're back to a duel between Ugly and Bad, but this time Ugly gets the first shot. His chances of winning are a bit better than ⅓ – in fact, they are ³⁄₇ or 43 per cent. If Bad *misses* Good, then Good kills Bad next shot and we are left with a duel between Ugly and Good, with Ugly shooting first. Now his chances of survival are exactly ⅓.

In other words, Ugly's prognosis is much better when he misses either opponent rather than killing them. He must therefore miss at all costs. His best course of action is to shoot into the air.

In fact, by missing both, he has the best chance of survival of the three. I won't bog you down with the probability calculation, but Ugly has about a 40 per cent chance of being the last one standing, Bad's chance is about 38 per cent, and Good has only about a 22 per cent chance.

The moral of the story is this: wherever possible, let the toughest fight between themselves.

⑬ APPLES AND ORANGES

There are three boxes – labelled 'apples', 'oranges' and 'apples and oranges' – and we must remove a piece of fruit from one of them.

Let's go through the possibilities when we choose a piece of fruit from each of the boxes. Say we take an item from 'apples'. If it is an apple, we know that the box must contain both apples and oranges. The box can't contain *only* apples because we know the labelling is wrong, and the label says 'apples'. This leaves us with two boxes left, 'oranges' and 'apples and oranges', and only two possible contents, one of only oranges and one of only apples. The box 'oranges' cannot have oranges, since the labelling is wrong, so it must have apples. This leaves 'apples and oranges' with oranges, and we have determined correctly the contents of the three boxes. Hurrah! It appears that we have solved the problem. Only we haven't. Since if our strategy was to choose a fruit from 'apples', there is also a chance that the fruit will be an orange. And if we take an orange from 'apples' then 'apples' is either oranges or apples and oranges, and we have no way of telling which. Likewise, say we choose an item from 'oranges'. There's a chance we get an apple, which means we cannot tell whether the box is apples, or apples and oranges.

The solution is that we have to choose a fruit from the box labelled 'apples and oranges'. In fact, you may have deduced this already without having to go through the workings of the previous paragraph. If in a puzzle there is a single solution from a choice of three, where two of the three choices are interchangeable, as 'apples' and 'oranges' are here, then the solution must be a consequence of choosing the odd one out.

So, we take a fruit from 'apples and oranges'. If it is an apple, we know that the box must contain just apples. This leaves us with the boxes labelled 'apples' and 'oranges', which are the box of oranges and the box of apples and oranges. The box labelled 'oranges' does not have just oranges in it, so it must be the box containing both fruit. So the box labelled 'apples' contains oranges. We are thus able to correctly relabel all the boxes. As we would have been able to if the fruit from 'apples and oranges' was an orange, since we would go through the same process but switching apples for oranges.

⑭ SALT, PEPPER AND RELISH

First we need to establish who the man is. Sid sounds a likely candidate. But this will lead us into a contradiction. The question tells us that the man doesn't have the relish. If Sid is the man, he can't have the salt either because of his surname, so he must have the pepper. Reese cannot have the relish because of his surname, but neither can he have the salt, because in the dialogue he responds to the person who does. So Reese must also have the pepper and we have a contradiction.

So is Phil the man? Phil's a manly name! This again gets us to a contradiction. From the dialogue the man is *not* the person with the salt. So if Phil is the man he cannot have the salt, and he can't have the pepper either, since that matches his surname. So he must have the relish. But we are told the man does not have the relish.

By a process of elimination Reese must be the man. The man doesn't have the salt, so Reese must have the pepper. Sid must therefore have the relish and Phil the salt.

If you're interested, Sid is a diminutive of Sidney, an increasingly common girl's name, and Phil a diminutive of Philippa.

⑮ ROCK, PAPER, SCISSORS

We can deduce how the game played out in the following way:

Let's consider the six times that Adam uses scissors. Since we know that there are no ties, each of these six scissors must correspond to a time when Eve throws rock or paper. She has two rocks and four papers, so we can deduce that every time she used either rock or paper Adam used scissors. Adam's scissors lose twice (against rock) and win four times (against paper). The aggregate score is Adam 4 Eve 2.

In the remaining four bouts, Eve uses scissors every time. Adam uses three rocks and one paper. The score here is Adam 3 Eve 1.

The final tally is Adam 7 Eve 3.

Adam wins.

⑰ SOOT'S YOU

Miss Atkinson assumes her face is clean, and that the other two passengers are laughing at each other. (Let's say one is on the left, and one on the right). Miss Atkinson then puts herself in the mind of one of the passengers, say the one on the left. He can see the passenger on the right, who has a sooty face, and Miss Atkinson, who doesn't. So the passenger on the left is chuckling because the one on the right has soot on his face. But why – thinks Miss Atkinson – does the passenger on the left think that the passenger on the right is laughing? The one on the left assumes that he has no soot on his face, so who can the one on the right be laughing at? The only possibility is the unpleasant reality that he must be laughing at Miss Atkinson! She instantly feels for her handkerchief and cleans her face.

⑱ FORTY UNFAITHFUL HUSBANDS

If you have solved the last two problems – or at least read through the solutions – you are almost all the way there. You may have spotted that they are variations on a theme: the first involves two girls, the second three passengers and this one 40 wives.

In fact, if you increase the children in the muddy face puzzle from two to 40, replace 'has a muddy face' with 'has an unfaithful husband', and replace 'steps forward' with 'kills her husband', that puzzle is the same as this one.

What is truly brilliant here is that the information from an observer that 'at least one husband has been unfaithful' seems so inconsequential, and possibly redundant, to what comes next, because every woman knows that at least one husband has been unfaithful. Actually, they all know 39 of the cads. Yet the information triggers a remarkable sequence of events.

The muddy face puzzle concluded with the two girls realising they both had mud on their faces and stepping forward. The climax of this puzzle is a horror flick: a finale in which the 40 wives slaughter their husbands simultaneously

So how do we get there? Imagine what happens if only one husband is cheating on his wife, and all of the other 39 husbands are faithful. The wife of the solitary adulterer, of course, is not aware that there are any adulterers in the town, since all women start with the assumption that their husbands are faithful, and in this case she knows that all the other husbands are faithful too. The day she hears that at least one husband has been unfaithful, she realises that it must be her husband – since it can be no one else – and she kills him the following day at noon.

Now imagine two are love rats. These men's wives – let's call them Agnes and Berta – are both only aware of one unfaithful man, since wives are ignorant of their own husbands' infidelities. Agnes knows Berta's husband is unfaithful, and Berta knows Agnes's is. The other 38 wives know that both Agnes and Berta's husbands are. Since everyone is aware of at least one adulterer, news that 'at least one husband has been unfaithful' troubles no one in the town and the following day passes without bloodshed.

In the afternoon of that day, however, both Agnes and Berta are confused. Agnes has deduced – as we did above – that if Berta's husband is the only adulterer in the town then Berta should have killed him at noon the day after she heard that at least one husband has been unfaithful. The fact Berta did not kill him leads Agnes to the conclusion that Berta must know of a second adulterer. And who could he be? He can't be anyone's husband except her own! So the following day Agnes kills her husband at noon, at the same time that Berta – who made the same realisation – kills her husband too. In other words, when there are two unfaithful husbands, on the second day after it is announced that 'at least one husband has been unfaithful' they will both be killed.

We can continue with the scenario of three unfaithful husbands. Each of their wives will only think there are two unfaithful husbands, and so when the second day passes with all men still alive, the penny drops. These three women kill their husbands on the third day. Cut to the chase. If there are 40 unfaithful husbands, nothing happens until a bloodbath on the fortieth day.

Had the monarch omitted to mention that there was at least one adulterer, the logical argument above cannot be made and the massacre in the town square would be averted.

⑲ BOX OF HATS

The only way that Algernon can know the colour of his hat is if he sees two green hats on the others, meaning that he has red. If he doesn't know the colour of his hat, therefore, he must see either two red hats or a red and a green.

Likewise, Balthazar must also see either two red hats, or a red and a green. We don't seem to be making much progress, however, since all we have established is that either [1] everyone has a red hat; [2] Algernon and Balthazar have green hats; or [3] only Caractacus has a green hat.

When we are told that Caractacus sees only red hats, we can eliminate [2]. Now imagine that [3] is true: that Caractacus has a green hat. If this were true, then let's go through the question again. Algernon would see a green hat and a red hat, and conclude that he does not know the colour of his own hat. Balthazar can see that Caractacus has a green hat. Since Algernon doesn't know the colour of his hat, Balthazar can rule out having a green hat himself, because if he did

Algernon would have said he knew the colour of his own hat! So Balthazar knows he has a red hat, in which case he would *not* be able to say that he doesn't know the colour of his hat. If we assume [3] is true we are led into a contradiction, so [1] is true: Caractacus has a red hat.

(20) CONSECUTIVE NUMBERS

To solve this puzzle we take some information from each of the statements, gradually reducing the set of potential numbers that Zebedee could have chosen.

The numbers available are 1, 2, 3, 4, 5, ... If Xanthe has 1, then since she knows the numbers are consecutive, she knows that Yvette must have 2. So Xanthe cannot have 1, and we can strike 1 from the list. If, on the other hand, Xanthe has 2, then Yvette could have 1 or 3, so she would not know Yvette's number. Likewise, for all numbers greater than 2, there is always the possibility that Yvette has one less or one more than that number. So all we know from the first statement is that Xanthe's number is 2 or greater.

Yvette cannot have 1 for the same reason as above. But can she have 2? If she has 2 then she knows that Xanthe must have either 1 or 3. But as she is a perfect logician she has deduced that Xanthe does not have 1. So if Yvette has 2, she knows Xanthe must have 3, but that contradicts her statement that she does not know Xanthe's number. We can delete 2 from Yvette's list. If Yvette has 3 or above, then she is truthful in saying she does not know Xanthe's number, since Xanthe can logically have Yvette's number plus or minus one.

To recap, we know that Xanthe has one of 2, 3, 4, 5, 6, ... , and that Yvette has one of 3, 4, 5, 6, 7, ...

Xanthe now says that she knows the number. If she has 2, then she knows that Yvette has 3. If she has 3, then she knows that Yvette must have 4. If Xanthe has 4, Yvette could have 3 or 5, so she does not know the number. And likewise for numbers greater than 4. In other words, for Xanthe to truthfully say she knows Yvette's number she must have either 2 or 3.

If Xanthe has 2 or 3, Yvette must have either 3 or 4, since Zebedee's numbers are consecutive. So Zebedee whispered either 2 and 3, or 3 and 4, to the girls. We can conclude that one of his numbers was definitely 3.

(21) CHERYL'S BIRTHDAY

Cheryl lists her potential birthdays and then tells Albert the month, which is either May, June, July or August. She tells Bernard the day, which is either 14, 15, 16, 17, 18 or 19. From each line of dialogue we will gather information that allows us to eliminate certain months or days, and by the end all the options will be eliminated except the answer.

Albert says that he doesn't know when Cheryl's birthday is, but that he also knows Bernard doesn't know.

Every month appears at least twice in Cheryl's list, so whichever month she told Albert, there are a choice of at least two possible birthdays for it. So of course Albert doesn't know her birthday. The first part of the sentence is redundant.

For Albert to know that Bernard doesn't know, however, he must know that Bernard cannot have a number that appears only once on the list. Those numbers are 18 and 19. If Bernard had been told either 18 or 19 then he would be able to deduce Cheryl's birthday. And the only way for Albert to know that Bernard does not have 18 or 19 is if Cheryl told him a month with no 18 or 19 options. So, we can eliminate the months with 18 and 19, which are May and June. Albert must have been told July or August.

Bernard says that at first he didn't know Cheryl's birthday, confirming he doesn't have 18 or 19 – but then he says he knows. To be able to say this, he must have a number for which there is only one remaining choice of month. We can therefore eliminate 14, since both 14 July and 14 August are options. So Bernard must have 15, 16 or 17.

Albert now says he knows the birthday, so he must have a month for which there is only one remaining choice. The only dates left are 16 July, 15 August and 17 August. So, the answer is 16 July.

Even though 16 July is the correct answer, however, a subtle difference in the interpretation of the question leads to a different answer, 17 August. The online debate about which is the right approach probably helped the puzzle to be shared and talked about. To quell the controversy, Singapore and Asian Schools Math Olympiad, which set the question, made a public announcement clarifying the question and stating that 17 August was wrong.

Here's how you get the alternative date. It's a great case study in just how nuanced logic puzzles can often be, and how careful you need to be about what is assumed knowledge and what is not.

Albert starts off by saying that he does not know Cheryl's birthday, but that he *knows* Bernard doesn't know. If you assume that he knows this information about Bernard because he deduced it, then, as above, we end up with 16 July as the answer. But maybe Albert knows of Bernard's ignorance because he has been told of it beforehand.

If this is so, Albert can eliminate the dates with 18 and 19, which are the numbers that appear only once, before he starts. When Albert then says he doesn't know the birthday, he is revealing that he does not have June, since that month only appears once in the remaining dates – so we can eliminate June. But, unlike the previous scenario, he still might have May, so we cannot eliminate it. The dialogue continues, with Bernard saying he didn't know the birthday but now he does. For him to know, he must have a number that appears only once among the remaining possibilities: 15 and 16 May, 14 and 16 July, and 14, 15 and 17 August. The number 17 appears only once, so the answer is 17 August.

All very logical, but I agree with the Singapore and Asian Schools Math Olympiad that the most natural interpretation of the question is that Albert *deduces* that Bernard doesn't know Cheryl's birthday, not that he is told this beforehand.

㉒ DENISE'S BIRTHDAY

You solve this in the same way as the Cheryl problem. Each sentence gives you a clue as to what to eliminate. Yet the Denise problem is more involved, and pushes your brain to think in different directions.

Here are the dates again:

17 February 2001	16 March 2002	13 January 2003	19 January 2004
13 March 2001	15 April 2002	16 February 2003	18 February 2004
13 April 2001	14 May 2002	14 March 2003	19 May 2004
15 May 2001	12 June 2002	11 April 2003	14 July 2004
17 June 2001	16 August 2002	16 July 2003	18 August 2004

Albert (who has the month) knows that Bernard (who has the day) doesn't know. The only days that appear exactly once in the suggested dates are 11 and 12, so we can eliminate those dates, which are 11 April 2003 and 12 June 2002. Cross them out from the table if that's helpful.

Since Albert knows that the month is not April or June, we can delete all the other April and June dates: 13 April 2001, 15 April 2002 and 17 June 2001.

Bernard (who has the day) still doesn't know Denise's birthday, so we can eliminate all remaining dates with numbers that appear only once, since if he had one of these dates he would know the birthday. Both 15 and 17 appear only once, so we can cross out 15 May 2001 and 17 February 2001.

But Bernard also knows that Cheryl (who has the year) doesn't know Denise's birthday. The only way Cheryl could know is if she has 2001, since there is only one option left for that year, 13 March 2001. So Bernard does not have 13, and we can eliminate it from the grid. Goodbye 13 March 2001 and 13 January 2003.

The fact that Cheryl does not know when the birthday is does not give us new information, but if she knows that Albert still doesn't know, then Albert cannot have a month that appears only once in the remaining options. The only remaining month that appears once is January, in 19 January 2004, which means that Cheryl does not have 2004. We can delete all the 2004 dates.

Albert now knows the date, so the month must appear only once in the remaining options. So we can delete the two March dates, leaving us with 14 May 2002, 16 August 2002, 16 February 2003 and 16 July 2003.

If Bernard now knows, the day must appear only once in the remaining dates. The answer is 14 May 2002.

(23) THE AGES OF THE CHILDREN

The verger has three children. When you multiply their ages together you get 36. This piece of information means we can reduce the number of possible ages to the combinations below. The final column in bold is the sum of the three ages.

1	×	1	×	36	**38**
1	×	2	×	18	**21**
1	×	4	×	9	**14**
1	×	6	×	6	**13**
2	×	2	×	9	**13**
2	×	3	×	6	**11**
3	×	1	×	12	**16**
3	×	3	×	4	**10**

We can assume that the vicar knows, or can find out, the number on the verger's door. If the door number was the same as a number that appears only once in the bold column, the vicar would immediately know the ages of the children. But if the door number was 13, he would require more information. So we deduce that the door number is 13 and the ages are either:

1, 6, 6 or 2, 2, 9

The vicar obviously knows the age of his son, and we can assume that the verger does too. Since the verger has told the vicar that this piece of information is enough to determine the ages, the vicar's son must be older than all the children in one set of possible ages, but younger than at least one child in the other set of possible ages. In other words, the vicar's son *must* be either 7 or 8. If the vicar's son was, say, 10 or 11, then since the son would be older than all the children in both possibilities, the verger would *not* have been able to state that the vicar could solve the problem. And if the vicar's son is 7 or 8, the verger's children are 1, 6 and 6.

24 WIZARDS ON A BUS

We know that wizard A has more than one child, that the ages of his children are positive whole numbers and that these ages add up to the bus number. With this information, let's test different bus numbers.

The bus number cannot be 1, since there are no two positive whole numbers that add up to 1.

If the bus number is 2, the children must be 1 and 1, the only positive whole numbers that add up to 2. This means wizard A is also aged 1, since his age is the product of his kids' ages. Ignoring biology, we can also eliminate this possibility using logic. Wizard A told wizard B that knowing both his age and the number of his children would not allow wizard B to deduce his children's ages. But here, if wizard A is aged 1 and has two children, then wizard B would be able to deduce the children's ages: 1 and 1, which multiply to 1. So the bus number cannot be 2.

If the bus number is 3, there are two possibilities: three children aged 1, 1, and 1 (giving the wizard an age of 1); or two children aged 2 and 1 (giving the wizard an age of 2). But in either case, wizard B would still be able to deduce the children's ages. For example, if wizard A is aged 2 and has two children, their ages must be 2 and 1, the only positive whole numbers that multiply to 2. We can now see the pattern. We are looking for a case where, for a certain combination of wizard A's age and the number of children, there is more than one possible set of ages for the children – which would prevent wizard B from deducing them.

Let's carry on with bus number 4. Here is a table containing possible ages of the children that add up to 4, the number of those children, and the age of the wizard:

Ages of the children	Number of children	Age of wizard
3, 1	2	3
2, 2	2	4
2, 1, 1	3	2
1, 1, 1, 1	4	1

No two rows in this table share the same values in column two *and* column three. This means that every possible combination of 'number of children' and 'age of wizard' is tied to a unique set of ages in column one. In every case, if given the other values, wizard B will be able to deduce the children's ages – so the bus number cannot be 4.

You may need several sheets of paper to continue inspecting bus numbers in this way. But eventually you will get to bus number 12.

Here's a partial table:

Ages of the children	Number of Children	Age of wizard	Ages of the children	Number of Children	Age of wizard
11, 1	2	11	5, 2, 2, 2, 1	5	40
10, 2	2	20	5, 2, 2, 1, 1, 1	6	20
10, 1, 1	3	10	5, 2, 1, 1, 1, 1, 1	7	10
9, 3	2	27	5, 1, 1, 1, 1, 1, 1, 1	8	5
9, 2, 1	3	18	4, 4, 4	3	64
9, 1, 1, 1	4	9	**4, 4, 3, 1**	**4**	**48**
8, 4	2	32	4, 4, 2, 2	4	64
8, 3, 1	3	24	4, 4, 2, 1, 1	5	32
8, 2, 2	3	32	4, 4, 1, 1, 1, 1	6	16
8, 2, 1, 1	4	16	4, 3, 3, 2	4	72
8, 1, 1, 1, 1	5	8	4, 3, 3, 1, 1	5	36
7, 5	2	35	4, 3, 2, 2, 1	5	48
7, 4, 1	3	28	4, 3, 2, 1, 1, 1	6	24
7, 3, 2	3	42	4, 3, 1, 1, 1, 1, 1	7	12
7, 3, 1, 1	4	21	4, 2, 2, 2, 2	5	64
7, 2, 2, 1	4	28	4, 2, 2, 2, 1, 1	6	32
7, 2, 1, 1, 1	5	14	4, 2, 2, 1, 1, 1, 1	7	16
7, 1, 1, 1, 1, 1	6	7	4, 2, 1, 1, 1, 1, 1, 1	8	8
6, 6	2	12	4, 1, 1, 1, 1, 1, 1, 1, 1	9	4
6, 5, 1	3	30	3, 3, 3, 3	4	81
6, 4, 2	3	48	3, 3, 3, 2, 1	5	54
6, 4, 1, 1	4	24	3, 3, 3, 1, 1, 1	6	27
6, 3, 3	3	54	3, 3, 2, 2, 2	5	72
6, 3, 2, 1	4	36	3, 3, 2, 2, 1, 1	6	36
6, 3, 1, 1, 1	5	18	3, 3, 2, 1, 1, 1, 1	7	18
6, 2, 2, 2	**4**	**48**	3, 3, 1, 1, 1, 1, 1, 1	8	9
6, 2, 2, 1, 1	5	24	3, 2, 2, 2, 2, 1	6	48
6, 2, 1, 1, 1, 1	6	12	3, 2, 2, 2, 1, 1, 1	7	24
6, 1, 1, 1, 1, 1, 1	7	6	3, 2, 2, 1, 1, 1, 1, 1	8	12
5, 5, 2	3	50	3, 2, 1, 1, 1, 1, 1, 1, 1	9	6
5, 5, 1, 1	4	25	3, 1, 1, 1, 1, 1, 1, 1, 1, 1	10	3
5, 4, 3	3	60	2, 2, 2, 2, 2, 2	6	64
5, 4, 2, 1	4	40	2, 2, 2, 2, 2, 1, 1	7	32
5, 4, 1, 1, 1	5	20	2, 2, 2, 2, 1, 1, 1, 1	8	16
5, 3, 3, 1	4	45	2, 2, 2, 1, 1, 1, 1, 1, 1	9	8
5, 3, 2, 2	4	60	2, 2, 1, 1, 1, 1, 1, 1, 1, 1	10	4
5, 3, 2, 1, 1	5	30	2, 1, 1, 1, 1, 1, 1, 1, 1, 1, 1	11	2
5, 3, 1, 1, 1, 1	6	15	1, 1, 1, 1, 1, 1, 1, 1, 1, 1, 1, 1	12	1

At last, we have found what we were looking for (in bold): two rows with identical values for both 'number of children' and 'age of wizard'. If wizard A were to tell wizard B that he was 48 years old and had four children, wizard B would *not* know enough to deduce the children's ages. They could be (6, 2, 2, 2) or (4, 4, 3, 1). The bus number is 12.

I told you there was only one possible bus number, so you could stop here. But you may enjoy seeing why this is true. Consider bus number 13. If you drew up a table, you would find two ways for wizard A to be 48 with five children: (6, 2, 2, 2, 1) or (4, 4, 3, 1, 1). You would also find two ways for him to be 36 with three children: (6, 6, 1) or (4, 3, 3). So, once wizard A tells wizard B that his age and number of children is insufficient for wizard B to work out the kids' ages, wizard B will deduce that wizard A is 48 or 36—but he cannot know which is correct, and so he cannot say 'Aha! AT LAST I know how old you are!' as he does.

By adding a one-year-old child to each of the above groupings for bus 13, we can create the following possible family scenarios for bus 14: either (6, 2, 2, 2, 1, 1) and (4, 4, 3, 1, 1, 1) with wizard age 48; and either (6, 6, 1, 1) or (4, 3, 3, 1), with wizard age 36. In this case again, wizard B cannot deduce wizard A's age. (Each time we add an extra 1, the product does not change, so wizard A's age stays the same, and the number of children increases by the same amount for each grouping.) Thus, we eliminate bus 14 – and indeed, by adding more one-year-olds, we can eliminate 15, 16, and all other numbers.

Conway's genius is discovering that the question's delicious set-up provides only one possible bus number.

(25) VOWEL PLAY

You need to turn over the A card and the 2 card.

Clearly the A card needs to be turned over, because you need to check there is an odd number on the back. We do not need to turn the B card over, because B is a consonant and we don't care about those.

The error most people make is in thinking that the 1 card needs to be turned over to check there's a vowel on the other side, since 1 is odd. But this logic is fallacious. If the other side is a vowel, the rule is verified. If the other side is a consonant, however, it doesn't matter what number appears face up, because the rule does not concern consonants.

However, we do need to turn over the 2, to make sure that there is no vowel on the other side, because if there was the rule would be broken.

Psychologist Peter Wason came up with this puzzle in 1966. The reason most people get it wrong is not that they don't understand the question. Rather they fall into the trap of reasoning from what they do know – the odd number facing up – instead of from what they don't know – what's face down. Our lazy brains are not designed to solve logic puzzles!

However, when the same puzzle is phrased slightly differently, using a familiar social context, most people get the right answer. The following cards all have a drink on one side and a number on the other. Each card represents a person. The number is their age, and the drink is their particular tipple.

To verify the following rule:

If a person is drinking alcohol, they are over 18.

Which cards do you need to turn over?

Obviously we need to turn over the wine card. But it is much clearer that we need to turn over 17 to see what they are drinking. We don't need to find out what the 22-year-old is drinking because they can drink what they like.

Ten tasty teasers

Are you a wizard at wordplay?

1) **S**LYLY

2) TYPEWRITER: the question asked for a word that uses *only* the ten keys, not that uses *every one* of them.

3) **T**ONIG**HT**

4) June July August September October November December January February March April May

5) **EXT**RAOR**DINARY**

6) F for forty: the sequence is the first letters of the numbers from seventeen to thirty-nine.

7) **EART**H**QUAKE**

8) Remove the first letter and each of the remaining letters form palindromes, meaning they are the same forwards and backwards:

   ssess, anana, resser, rammar, otato, evive, neven, oodoo

9) **INS**TANTAN**EOUS**

10) U. Whenever there are seven in a sequence, think days of the week: **M**onday, **T**uesday, **W**ednesday, **Th**ursday, **F**riday, **S**aturday, **Su**nday.

A Man Walks Round an Atom...

GEOMETRY PROBLEMS

(26) THE LONE RULER

A ruler enables us to draw straight lines. A ruler with a 2-unit interval enables us to draw straight lines *and* mark out intervals of 2 units. That's all we have to work with, but it's enough.

Our solution is based on the principle that two straight lines starting at the same point will diverge at a constant rate. We are looking to find a way to measure the distance between two divergent lines that allows us to create another line that is only 1 unit long. Here's how we do it.

Step 1. Draw two lines that cross each other. These are our divergent lines. Mark the points that are two units from the intersection down both lines. Using these new points, mark the points that are two units further down.

Step 2. Join the first two points and the second two points with a line. These lines are parallel. Mark point X, which is two units along the bottom line.

Step 3. Draw a line from the intersection to X. Mark the point Y where this line crosses the upper parallel line. The distance along the parallel line to Y, marked in bold below, is 1 unit. We have our answer.

Here's why it works. I've called one of the original lines A and the last line we drew B. The distance from A to B at their intersection is zero. As you move steadily down B, the distance from A to B along any line at a fixed angle increases at a constant rate. So, if a line from A to B at X has length 2, then the parallel line from A to B at Y, which is half the way there, must have length 1.

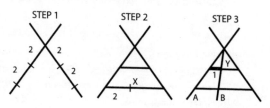

㉗ ROPE AROUND THE EARTH (NEW PROBLEM)

The rope can be lifted about 120m, which is about the height of the Centre Point skyscraper in central London. Again, the distance seems counter-intuitively huge. We have a 40,000km rope around the Earth, and by increasing it to 40,000.001km it lifts up so high that a pyramid of giraffes driving on motorbikes and stacked 20 giraffes high would easily ride through.

This time, though, the size of the Earth is very much relevant to the answer. The calculation requires some trigonometry, which will be too arcane for most readers, so full marks if you illustrated the problem correctly and *thought* about how you might solve it. In the diagram here, r is the radius of the Earth and h is what we are after, which is the height of the rope when raised as high as possible without stretching. The length of the rope from its peak to the ground is t, and the distance along the ground when the rope is in the air is twice g.

We can find h in terms of r, but be warned – it's not pretty. And don't even think about trying to follow if you never studied trigonometry. First, notice that t is a *tangent* to a radius, so we have a right-angled triangle in which the hypotenuse is $r + h$ and the other sides are r and t. Using Pythagoras's theorem:

[1] $t^2 + r^2 = (r + h)^2$

We know that the cosine of angle θ is $r/(r +h)$, so:

[2] $\theta = \cos^{-1} r/(r +h)$

And, since we are using radians, $\theta r = g$:

[3] $r \cos^{-1} r/(r +h) = g$

We also know from the statement of the question that:

[4] $2g + 1 = 2t$

These equations can be rearranged and 'simplified' – they can, believe me – to the equation: $h \approx (\frac{1}{2})(\frac{3}{2})^{(\frac{2}{3})}r^{(\frac{1}{3})}$

And when $r = 6,400,000$m, $h \approx 122$m

There we are. I showed it for completeness, and I promise there is no more trigonometry in this book.

㉘ BUNTING FOR THE STREET PARTY

The purpose of this question, like the previous one, is to confront our intuitions about space. The pole will be just over 7m high, about the height of a Victorian two-bedroom house, and way taller than the world's tallest giraffe. Surprisingly high, right?

Pythagoras gets us there without breaking a sweat. As illustrated here, the pole produces two right-angled triangles.

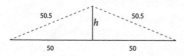

Each side of the bunting is a hypotenuse, and the ground and the pole are the other two sides. So:

$h^2 + 50^2 = 50.5^2$

$h^2 + 2,500 = 2,550.25$

Rearranging, we get:

$h^2 = 2,550.25 - 2,500 = 50.25$

So:

$h = \sqrt{50.25} = 7.1$

㉙ ON YER BIKE, SHERLOCK!

In order to work out which way the bike is travelling we first need to work out which of the tracks is the front wheel and which is the back. For us to be able to make this deduction, dear Watson, we need to understand how the curve of a tyre track determines the position of the wheel.

If a tyre track is straight, the wheel making that track is in line with the direction of the track. If a tyre track is curved, however, the wheel making that track is in line with the *tangent* of the curve at every point along the track. (The tangent is the line that touches a curve only at a single point). To see what I mean, consider the track opposite, made by a unicycle. When the wheel was at points A, B and C, it was in line with the tangents, which I've marked.

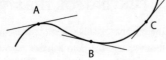

Bicycles have two wheels. The front wheel is free to point in any direction. But the back wheel has no freedom of direction – it must *always* be pointing in the direction of the front wheel.

Wherever the back wheel is, therefore, the front wheel is exactly one bicycle-length ahead of it in the direction of the tangent. In other words, all the tangents from the back wheel's track must cross the front wheel's track, and they must all do so one bicycle-length away.

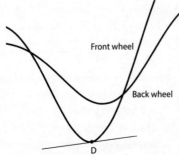

Now look at point D opposite on the bold line. Its tangent does not cross any track at all. We can deduce, therefore, that D does not sit on the back wheel track. It sits on the front wheel track.

Finally, we can find the bike's direction. We know which track was made by the back wheel, and we know from above that one bike-length along any tangent on that track in the direction of travel is a point on the front wheel track. So all we need to do is follow the tangent segments in both directions from E and F and see where they cut the front wheel track. The segments from E and F towards the left are of equal length, but the segments from E and F towards the right are not equal. Since the distance between wheels does not change during travel, the bike was going from right to left. Elementary.

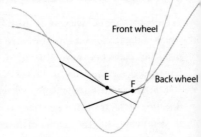

㉚ FUZZY MATH

I wheelie love this question because it illustrates a curious phenomenon: the top of a wheel always travels faster than the bottom.

When a wheel rolls along a horizontal surface, the points on the wheel are subject to two different directions of motion: they are moving horizontally, in the direction of travel, but also rotationally, around the centre of the wheel. These two directions of motion combine, and sometimes cancel each other out. Consider a point on the edge of a wheel. When that point is at the top, point A below, both its horizontal and its rotational motion are complementary. But when that point is at the bottom, point B, its horizontal and rotational motion are in opposite directions, and cancel each other out. From the perspective of someone watching, the point at the top of a rolling wheel is always travelling at twice the horizontal speed of the wheel, and the point at the bottom is always stationary. It follows that the points on the bottom half of the wheel are moving slower than the points on the top half.

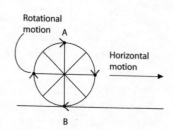

The answer, therefore, is the second image, where the pentagon is blurred above but sharp below. This would happen when the exposure time of the photographer's camera is short enough to get a crisp image of the slower pentagon but not the faster one. If you are an artist you may have got this instantly – the tops of moving wheels are often drawn as a blur.

㉛ ROUND IN CIRCLES

The answer that you probably got was (b) 3, which is what the examiners thought was the correct answer.

The calculation they wanted the students to do was as follows. If the radius of A is a third of the radius of B, then the circumference of A is a third of the circumference of B (since the circumference is 2π times the radius). You can therefore fit three circumferences of A around a single circumference of B. When A rolls for a full rotation it covers a single circumference. So it must roll for three full rotations to complete three circumferences, which is the circumference of B.

Their mistake is hard to spot unless you have studied the behaviour of circles rolling around circles. The examiners evidently hadn't. But let's do it now. Take two identical coins and roll one around the other. The coins have the same circumferences, so you might expect (like in the SAT question) the rolling coin to revolve only once before it gets back to its starting point. Yet the queen's head will revolve *twice*! When a circle rolls around a circle you need to add an

extra rotation, to account for the fact that it is rolling around itself *and* the other circle.

Had the SAT question been 'How many times will A rotate *along a straight line* of length equal to the circumference of B?', the answer would have been three. But when A rotates *around* B the answer is four.

The correct solution was not one of the multiple choice answers, which explains why almost no one got it right. The repercussions were embarrassing for the examiners: when their mistake was discovered the story appeared in the *New York Times* and the *Washington Post*.

(32) EIGHT NEAT SHEETS

The sheet directly underneath 1 can only be the sheet in the top left corner. The sheet directly underneath the top left corner has to be the one below it. And so on as the sheets spiral anticlockwise.

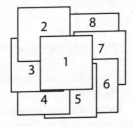

㉝ A SQUARE OF TWO HALVES

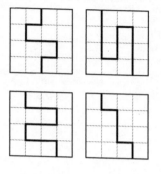

㉞ THE WING AND THE LENS

This problem becomes easier to understand if we pan out. Place four identical quarter circles together to get a big circle that comprises four smaller circles that overlap.

If r is the radius of the big circle, then the area of the big circle is πr^2.

The smaller circles have half the radius of the big one, so the area of each small circle is $\pi(\frac{r}{2})^2 = \frac{\pi r^2}{4}$.

This is neat. The small circles each have exactly a quarter of the area of the big one, so the area of four small circles is equal to the area of the big one. This equivalence in area is extremely helpful to us, since the image includes four small circles.

The small circles overlap. What is the total area of these four overlapping circles?

The area is equal to the area of four small circles (πr^2) minus the area of the overlap, which is the lenses.

[1] Overlapping circles area = πr^2 – area of lenses

We can also see that the area of the big circle (πr^2) minus the wings is equal to the overlapping circles.

[2] Overlapping circles area = πr^2 – area of wings

Combining these two equations, we get:

πr^2 – area of lenses = πr^2 – area of wings

Clearly, then, the area of the lenses is equal to the area of the wings. Since there are four equally sized wings and four equally sized lenses, the area of a single wing equals the area of a single lens.

(35) SANGAKU CIRCLES

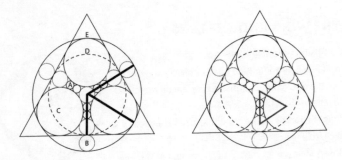

The perfect way in which all the circles fit so snugly together is what makes the image so alluring – and it is also the key to solving the puzzle, since it enables us to compare the circles' radii.

Let us call the circles, from smallest to largest, A, B, C, D and E, and let their radii be a, b, c, d and e. The question asks us to express d in terms of a.

In the first image, I have marked three lines. The vertical one is the radius of D, the dashed circle, but it also corresponds to four radii of A and three of B. So we can write the equation:

[1] $d = 4a + 3b$

Likewise, the other two thick lines, both radii of E, can be described in terms of other radii:

[2] $e = 4a + 5b$

[3] $e = b + 2c$

The clever bit now is to realise that the triangle (in the second image) is equilateral. The angle at the centre of E must be 60 degrees, and two of the sides are equal, so the third side must also be equal. In other words:

[4] $4a + 2b = b + c$.

We have four equations with five unknowns. Since we want to know d in terms of a let's get rid of the other terms.

First, we can lose e by equating [2] and [3].

$4a + 5b = b + 2c$

So:

$4a + 4b = 2c$, or:

[5] $2a + 2b = c$

Substituting c in [4] gives us:

$4a + 2b = b + 2a + 2b$, or:

[6] $2a = b$

And substituting in [1] gives us:

$d = 4a + 6a = 10a$

We have the answer: the radius of D is ten times the radius of A.

(36) SANGAKU TRIANGLE

I've renamed the circle sizes A, B and C, and their radii a, b and c. Our strategy will be first to find b in terms of a, and then c in terms of b, which will allow us to show that $c = 2a$.

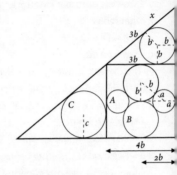

In the illustration on the right I've drawn a triangle with a dotted line. The length of the hypotenuse is $b + a$, because it covers a radius of each circle, and the lengths of the other two sides are b and $2b - a$. The second length you can deduce by realising that the triangle is half the length of the square – which must have length $4b$ – minus one radius of A.

Pythagoras's theorem tells us that for right-angled triangles the square of the hypotenuse is equal to the sum of the squares of the other two sides, so:

$(b + a)^2 = b^2 + (2b - a)^2$

Which expands to:

$b^2 + 2ab + a^2 = b^2 + 4b^2 - 4ab + a^2$

Which then contracts to:

$6ab = 4b^2$

And even smaller to:

$3a = 2b$

And finally to:

$b = \frac{3}{2} a$

So we have b in terms of a.

Now look at the top triangle in the illustration. I have drawn a line from the centre of the circle to each of the sides. Each of these lines meets the side at a right angle, so the triangle is split into a $b \times b$ square and two kite shapes. The long side of the kite pointing left is $3b$, since it is the width of the large square minus a radius of B. And since kites are symmetrical the other side of the kite must also be $3b$. If the side of the right-pointing kite is x, then using Pythagoras on the triangle we get:

$(3b + x)^2 = (b + x)^2 + (4b)^2$

Expanding, we get:

$9b^2 + 6bx + x^2 = b^2 + 2bx + x^2 + 16b^2$

Contracting, we get:

$4bx = 8b^2$

Which is:

$x = 2b$

The vertical side of the top triangle is $x + b = 2b + b = 3b$. The vertical side of the bottom triangle is $4b$. Since the two triangles are the same shape, although of different sizes, the ratio of the sides of the triangles, which is $\frac{3b}{4b} = \frac{3}{4}$, must equal the ratio of the radii of the circles inside the triangles, which is b/c.

If $\frac{3}{4} = \frac{b}{c}$

Then:

$c = \frac{4}{3} b$

We now have c in terms of b and b in terms of a.

So c in terms of a is $c = \frac{4}{3} b = \frac{4}{3} \left(\frac{3}{2}\right) a = 2a$.

(37) TREADING ON THE TATAMI

(38) FIFTEEN TATAMI MATS

This pattern is taken from the 1641 edition of *Jinkoki*, the most popular Japanese maths text book of the seventeenth century.

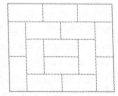

(39) NOB'S MATS

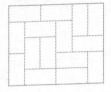

(40) AROUND THE STAIRCASES

No, tatami mats will not cover the 6 × 6 room with corners removed. If we shade in alternate squares like a chess board, as below, we can see why. Each mat must cover both a shaded square and a white square. So if the room is coverable by mats it must have an equal number of shaded and white squares. But this room has two extra white squares, so covering with mats is impossible.

Usually this question is phrased in terms of dominoes and a 'mutilated' chessboard – can dominoes the size of two chess squares tile a chessboard that has had opposite corners cut off? Again, for the same reason, the answer is no.

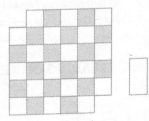

(41) RANDOM STAIRCASES

We show this with a clever technique devised by Ralph E. Gomory, who was director of research at IBM in the seventies. He was thinking about dominoes on a chessboard, but the proof is the same. First, draw a path through the room that visits every square just once, as shown overleaf. In the second image, I have removed one shaded and one white square arbitrarily for the staircases, which cuts the path into two sections. Each of the two sections of the path must cover an even number of squares, and can therefore be covered by tatami mats. The argument follows for all paths, and all choices of two differently coloured squares.

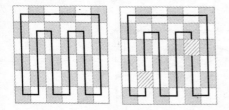

㊷ WOODBLOCK PUZZLE

This question was suggested to me by Joseph Yeo Boon Wooi, the Singaporean author of the Cheryl's Birthday problem (Problem 21), who had first read about it in the 1980s. The most obvious solution, illustrated in A below, is also what architects call a dormer window, or a vertical window that comes out of a sloping roof – as many of them were delighted to tell me. Two other solutions, B and C, are also possible.

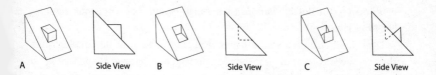

A   Side View   B   Side View   C   Side View

㊸ PICTURE ON THE WALL

We can solve this using physics (boo!) or maths (yay!). The former is predictably less elegant than the latter. Hammer two nails into the wall that are so close together that they will hold a piece of string wedged tightly between them. Then arrange the string like a W with the middle, upward tip of the W between the nails. The painting will hang as the nails will pinch the string in place, and when one of the nails comes off the painting will fall. Ugly, but possibly effective.

Here's a better solution.

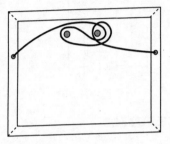

But this isn't my favourite way to string it up. What I was hoping you would do was to use the Borromean rings to reverse-engineer a solution, following my heavy hints that the Borromean rings provide a mathematical model of what we are trying to achieve. When one ring is removed, the other two become unlinked. The puzzle involves three elements – two nails and a piece of string – such that when one is removed they all become disconnected from each other. The hard part is working out just how two nails and a piece of string are equivalent to the Borromean rings, since neither nails nor string look anything like three rings.

Let's think again about the Borromean rings. They can be circular rings. They can be *valknut* triangles. They can, in fact, be whatever shape we like so long as the way they interlink is the same. Imagine, for example, that each nail is part of a rigid ring, perhaps one that goes from its tip through the wall, then up and round, back into the room and then joins the end of the nail again. And imagine that both ends of the string join up, making a giant loop around the room. If these three 'rings' are connected in a Borromean way then removing one nail will cause the string to become free from its loop round the other nail, giving us a solution.

So how do we it? I made myself a set of Borromean rings using two plastic rings and a piece of string, shown below left. I then separated the rings side by side (shown below right), as if they were the nails on the wall. The way the string loops between the rings is the solution we are after, which I have put below.

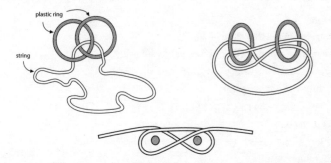

Note that the only section of each 'ring' that we are interested in is the bit representing the two nails and the string across the painting, since this is where all the interconnectedness is. The other bits of the 'rings' – the extension of the nails that go through the wall or the string that goes around the room – are irrelevant.

⑭ A NOTABLE NAPKIN RING

Let's finish off what I started. Since the height of the napkin ring is 6, half the height is 3. So the height of the dome, h, is equal to $r - 3$, as illustrated in the cross-section below.

To find a, the radius of the cylinder that's removed, we use Pythagoras's theorem on the right-angled triangle with the dotted line. The square of the hypotenuse is equal to the sum of the

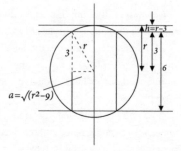

squares of the other two sides, so $r^2 = a^2 + 3^2$, and therefore $a = \sqrt{(r^2 - 9)}$.

Now it's time for the heavy lifting. Using the formulae for napkin volume, which we've established is:

sphere – cylinder – 2 × dome

We get:

$(\frac{4}{3})\pi r^3 - 6\pi a^2 - 2(\frac{\pi h}{6})(3a^2 + h^2)$

We replace a and h with our new expressions in r:

$(\frac{4}{3})\pi r^3 - 6\pi(r^2 - 9) - 2(\frac{\pi(r-3)}{6}(3(r^2 - 9) + (r-3)^2)$

Multiplying out, we get:

$(\frac{4}{3})\pi r^3 - 6\pi r^2 + 54\pi - (\frac{\pi(r-3)}{3})((3r^2 - 27) + (r^2 - 6r + 9))$

Keep on going …

$(\frac{4}{3})\pi r^3 - 6\pi r^2 + 54\pi - (\frac{\pi(r-3)}{3})(4r^2 - 6r - 18)$

Not long now …

$(\frac{4}{3})\pi r^3 - 6\pi r^2 + 54\pi - (\frac{\pi}{3})(4r^3 - 6r^2 - 18r - 12r^2 + 18r + 54)$

Sorry for the slog …

$(\frac{4}{3})\pi r^3 - 6\pi r^2 + 54\pi - (\frac{\pi}{3})(4r^3 - 18r^2 + 54)$

Almost there …

$(\frac{4}{3})\pi r^3 - 6\pi r^2 + 54\pi - (\frac{4}{3})\pi r^3 - 6\pi r^2 - 18\pi$

The terms in r now cancel, leaving:

36π

The answer is stunning. The term r does not appear in the answer, meaning that the size of the sphere is irrelevant to the question!

All napkin rings that are 6cm high have a volume of 36π. A 6cm-high napkin ring made by drilling through a sphere the size of an orange has the same volume as one made by drilling through a sphere the size of a beach ball, or even the Moon.

As you increase the circumference of the ring you make it thinner, and the increase in circumference and the decrease in thickness compensate each other perfectly at all sizes. Mind = blown.

⑤ AREA MAZE

Extend the image by adding the dotted lines. The size of area A plus the area sized 24cm² is equal to 9cm × 5cm, so A = 45cm² – 24cm² = 21cm². Meanwhile, A + B = 5cm × 8cm = 40cm². So B = 19cm².

B has the same width and the same area as the rectangle beneath it marked 19cm², so it must have the same height, and it must be identical to that rectangle. So A has the same height and width as the rectangle whose area we are looking for, so it must have the same area. The answer is thus 21cm².

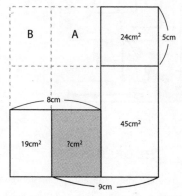

㊻ SHIKAKU

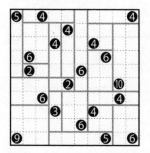

㊼ SLITHERLINK

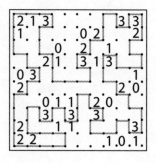

㊽ HERUGOLF   ㊾ AKARI

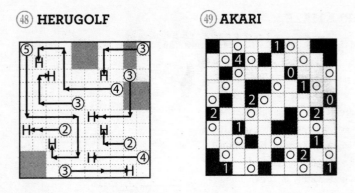

㊿ THE DARK ROOM

There are many solutions, all along the same lines. The room with the smallest number of walls has six of them, and looks like a three-spiked shuriken, the star-shaped weapon used by Japanese warriors. The square room is more architecturally realistic.

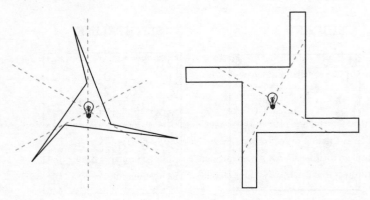

Ten tasty teasers

Are you smarter than a 12-year-old?

1) E

2) D

We can solve this is many ways. We could either give them all the common denominator 630: so, for example, $\frac{1}{2}$ is $\frac{315}{630}$, $\frac{2}{3}$ is $\frac{420}{630}$, and so on. Or we could write them out as decimal fractions. However, as they are all close in value to $\frac{1}{2}$, we could consider the value of each fraction minus $\frac{1}{2}$. Following the initial order, these values are 0, $\frac{1}{6}$, $\frac{1}{10}$, $\frac{1}{14}$ and $\frac{1}{18}$. So, when placed in order, the fractions are $\frac{1}{2}$, $\frac{5}{3}$, $\frac{7}{4}$, $\frac{9}{5}$ and $\frac{2}{3}$.

3) C

The letter 'e' already occurs eight times, so both 'nine' and 'eleven' may be placed in the gap to make the sentence true.

4) B

At an intersection of lines, the broken line is the one drawn first and the unbroken line is the one drawn over it. So we need to find a route for which the line is broken the first time it passes through any intersection and solid when it passes through that intersection for a second time. Only the route which starts at B and heads away from D satisfies this condition.

5) C

Of the options given, 23 × 34, 56 × 67 and 67 × 78 are not divisible by 5, so may be discounted. Also, 34 is not divisible by 4 and 45 is odd, so 34 × 45 may also be discounted, as it is not divisible by 4. The only other option is 45 × 56. If you break it down into its constituent prime numbers, $45 \times 56 = 2^3 \times 3^2 \times 5 \times 7$. It is clear that this number is divisible by all the whole numbers from 1 to 10 inclusive, since we have the primes 2, 3, 5 and 7 accounted for, and all the other numbers can be made up from this collection of primes: $4 = 2^2$, $6 = 2 \times 3$, $8 = 2^3$, and $9 = 3^2$.

6) B

If the Knave of Hearts is telling the truth, then the Knave of Clubs is lying, which means that the Knave of Diamonds is telling the truth, but the Knave of Spades is lying. Alternatively, if the Knave of Hearts is lying, then the Knave of Clubs is telling the truth, which means that the Knave of Diamonds is lying, but the Knave of Spades is telling the truth. In both cases we can determine that two of the Knaves are lying, although it's not possible to determine which two they are.

7) B

Consider each corner of the cube. Three faces meet there, and each pair of faces has an edge in common. So three different colours are needed. No other colours will be needed provided that opposite faces are painted in the same colour, since opposite faces have no edges in common.

8) B

If my current age is x, then Granny's age is $4x$. Five years ago: $4x - 5 = 5(x - 5)$. This equation reduces to $x = 20$. So Granny is 80 and I am 20.

9) B

We concentrate initially on the rightmost digits of the numbers given, which are 3, 5, 7 and 9, since these need to be added/subtracted to make a number ending in 0. We note that the 3 comes first, so is positive. Now 3 + 7 = 10, but there is no way to combine 5 and 9 to get a number ending in 0. So we must use 3 – 7. Hence, in the calculation, 67 must be preceded by a minus sign. Now 123 – 67 = 56. So we need to get an extra 44 by combining 45 and 89. The only way to do this is 89 – 45. So the correct calculation is 123 – 45 – 67 + 89. It has two minus signs and one plus sign, so $p - m$ is –1.

10) A

We can consider the tiling pattern to be a tessellation of the shape shown, so the required ratio is 1:1.

Mathematics Most Fowl

PRACTICAL PROBLEMS

(52) ONE HUNDRED BIRDS

As we did in the One Hundred Fowls problem, we need to turn this question into two equations, one for the number of birds, and one for the money. If the numbers of ducks, doves and hens are x, y and z, we have:

[1] $x + y + z = 100$

[2] $2x + y/2 + z/3 = 100$

Let's first multiply [2] by 6 so we get rid of the fractions.

[3] $12x + 3y + 2z = 600$

And let's multiply [1] by 2 so we get an equation with a $2z$ in it.

[4] $2x + 2y + 2z = 200$

Now we are all set to create a single equation by eliminating the $2z$ term. By rearranging [3] we have $2z = 600 - 12x - 3y$, and by substituting in [4] we have:

$2x + 2y + 600 - 12x - 3y = 200$.

This rearranges to:

[5] $10x + y = 400$

We know that x and y are whole numbers and that they are less than 100. We can also deduce something about y: it must be a multiple of 10. This is because 10 divides 400, so it must also divide the other side of the equation, $10x + y$. We also know that 10 divides $10x$. So we can deduce that 10 must also divide y, because if it didn't then 10 would not divide $10x + y$, which contradicts what we know.

The multiples of 10 that are less than 100 are 10, 20, 30, 40, 50, 60, 70, 80 and 90. But y can't be 70, 80 or 90, since when it has these values, x is 33, 32 and 31, so the sum of x and y, or ducks and doves, is more than 100. The six solutions are $y = 10$, 20, 30, 40, 50 and 60, which give the following divisions of birds:

ducks	doves	hens
39	10	51
38	20	42
37	30	33
36	40	24
35	50	15
34	60	6

(53) THE 7-ELEVEN

We are looking for the price of *four* items, but we are only given *two* statements about them: their sum and their product.

Let's write down the equations anyway. Let the prices of each item be a, b, c and d. The cashier tells us that:

[1] $a \times b \times c \times d = abcd = 7.11$

[2] $a + b + c + d = 7.11$

The fundamental theorem of arithmetic says that every whole number is equal to a unique set of prime numbers all multiplied together.

This theorem is going to be very useful, but we cannot apply it yet because it concerns whole numbers, and the multiplication in [1] includes a decimal number, 7.11. Our strategy is to turn [1] into an equation in whole numbers, and we do this by making the following substitutions.

Let $A = 100a$, $B = 100b$, $C = 100c$ and $D = 100d$. When we multiply them together, we get:

[3] $A \times B \times C \times D = ABCD = 100,000,000abcd$

But we know that $abcd = 7.11$, so

[4] $ABCD = 711,000,000$

Now we have a number we can work with. The fundamental theorem of arithmetic tells us that 711,000,000 has a unique set of prime factors, which are the primes that multiply together to make that number. You can work these out by hand, or (preferably) use a computer:

$711,000,000 = 2 \times 2 \times 2 \times 2 \times 2 \times 2 \times 3 \times 3 \times 5 \times 5 \times 5 \times 5 \times 5 \times 5 \times 79$

So:

$ABCD = 2 \times 2 \times 2 \times 2 \times 2 \times 2 \times 3 \times 3 \times 5 \times 5 \times 5 \times 5 \times 5 \times 5 \times 79$

The numbers A, B, C and D are therefore made up from these prime numbers. The question now is to work out which ones multiply together to make A, which ones multiply together to make B, which ones multiply together to make C and which ones multiply together to make D. In other words, how do we allocate these primes to A, B, C and D?

We now return to [2], and multiply it by 100 to get a second equation with A, B, C and D:

[5] $100a + 100b + 100c + 100d = A + B + C + D = 711$

In other words, we must allocate the prime numbers above among A, B, C and D such that A, B, C and D added together equals 711.

The bad news is that there are no shortcuts – it is just a case of using trial and error. For example, let's say that $A = 2 \times 2 \times 2 \times 2 \times 2 \times 2 = 64$, $B = 3 \times 3 = 9$, $C = 5 \times 5 \times 5 \times 5 \times 5 \times 5 = 15,625$, and $D = 79$. Then $A + B + C + D = 15,777$, so this doesn't work.

A good deal of the process is now down to luck, but slowly you will get a sense of roughly how big each of A, B, C and D need to be. And you can make educated guesses. There are lots of fives, so maybe two or three of the numbers will be multiples of five. If so, then the sum of these numbers will end in a 0 or a 5. Which means the final number must end in a 6 or a 1. What is the lowest multiple of 79 that ends in a 6 or a 1? It's 79×4, and sure enough:

$A = 79 \times 2 \times 2 = 316$

$B = 5 \times 5 \times 5 = 125$

$C = 5 \times 3 \times 2 \times 2 \times 2 = 120$

$D = 5 \times 5 \times 3 \times 2 = 150$

So, the prices a, b, c and d are £3.16, £1.25, £1.20 and £1.50.

The beauty of this puzzle is not in the rather laborious trial and error but in the brilliant way that the number 7.11 gives a unique solution for each of the four prices.

⑤④ THE THREE JUGS

Keep reading through the billiard table solution on pages 106–7.

⑤⑤ THE TWO BUCKETS

I hope you built that billiard table.

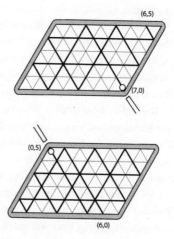

The first illustration shows what happens when you cue off from (7, 0), i.e., fill the seven-gallon bucket first. The second illustration cues off from (0, 5), i.e., filling the five-gallon bucket first. There are fewer rebounds in the first diagram before you hit an edge where the horizontal coordinate is 6, so this is the way to pour six gallons in the smallest number of pourings.

The rebound coordinates in the first illustration, which also represent the number of litres in each of the pourings, are: (7, 0), (2, 5), (2, 0), (0, 2), (7, 2), (4, 5), (4, 0), (0, 4), (7, 4) and (6, 5). So the quickest solution is to pour 7 into the first bucket with the second one empty, then to fill the second bucket with 5, leaving the first with 2, and so on, until you have 6 in the first bucket and the second bucket is full.

⑤⑥ THE WHITE COFFEE PROBLEM

Let's imagine there are 100ml of coffee in the flask and 100ml of milk in the bowl. And let's say we pour 10ml of coffee into the milk. There will now be 110ml in the ...

Stop right there!

Yes, we can solve this problem by taking arbitrary values, working it out, and then generalising – but there is a much quicker, more elegant way.

First, let's be clear that mixing the two liquids does not change the chemistry of either of them. The total volume of coffee and the total volume of milk never changes. What's not coffee in either receptacle is milk and what's not milk is coffee.

After both pourings, the volume in the flask is the same as it was at the beginning, but the flask now contains a volume of coffee molecules and a volume of milk molecules. Where has the lost volume of coffee gone? All of it is in the bowl, since the total amount of coffee has not changed. So, the volume of milk in the flask must be equal to the volume of coffee that is in the bowl. The size of the flask, the size of the bowl and the amount poured between them is irrelevant to the answer.

Perhaps it is easier to see this in terms of cookies and jars. There are chocolate cookies in one jar, and coconut cookies in another. Take an arbitrary amount of chocolate cookies and put them in the coconut cookie jar. Then remove the same amount of cookies from the coconut jar – which can now be made up of either type, since that jar contains a mix – and put them back in the chocolate jar.

The chocolate cookie jar now contains chocolate cookies and possibly some coconut ones. The number of coconut ones it does contain, however, is evidently the number of chocolate cookies left in the coconut jar.

57 WATER AND WINE

It is impossible for either jug to be half water and half wine if you are only pouring half pints between them – the only way to make this happen would be if you could pour the entire contents of one jug into the other.

Again, as in the problem above, do not try to solve this using numbers, as you will quickly be led into a quagmire of fractions ...

Think of it this way. When you pour from a jug with a higher concentration of wine into a jug with a lower concentration of wine, the first jug continues having a higher concentration of wine than the second jug. This is because the first jug continues with exactly the same concentration of wine, while the concentration of wine in the second jug will be somewhere between the original concentrations in both jugs.

Likewise, when you pour from a jug with a lower concentration of wine into a jug with a higher concentration of wine, the first jug continues having a lower concentration of wine than the second jug. This is because the first jug continues with exactly the same concentration of wine, while the concentration of wine in the second jug will be somewhere between the original concentrations in both jugs.

At the start one jug is 100 per cent wine, and the other 0 per cent wine. Since there is a difference in concentrations at the start, and each pouring – either from lower to higher, or vice versa – maintains at least some of the difference, then the two jugs can never be 50/50 water and wine.

⑤⑧ FAMOUS FOR 15 MINUTES

The 'aha!' moment here is the realisation that an hourglass can be flipped before it has fully emptied.

Let's flip both to start, as we did in the example, but this time once the 7-minute glass is empty, flip it again. After 11 minutes is up, the 7-minute glass will have been emptying for four minutes. Flip it for a third time! It will now take another four minutes to empty, and when it does the total time elapsed since the start is a quarter of an hour.

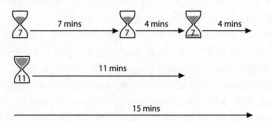

(59) A FUSE TO CONFUSE

[1] The question stated that if a fuse is cut in half there is no guarantee that either section will burn for half an hour. It follows that if you snip a quarter length off a fuse there is no guarantee that the remainder of the fuse will take 45 minutes to burn. The solution requires some lateral thought.

If you light a fuse at one end and then put it out after 30 minutes, the remaining section – however long it is – will take, once lit, 30 minutes to burn out. If you burn a fuse from *both* ends, it will always burn out in 30 minutes, even if it burns for a longer distance from one end than it burns from the other.

So, light one of the fuses at both ends, and the second fuse at only one end. After 30 minutes one fuse will have burnt out, and the other one will have 30 minutes left to go. At this moment light the other end of the second fuse. Lit at both ends it will die 15 minutes later, marking 45 minutes from the time it was first lit.

[2] When a single fuse is burning from one end it will burn out in an hour. When a single fuse is burning from both ends, it will burn out in half an hour. If a single fuse could burn from three ends, it would burn out in a third of an hour, or 20 minutes, since burning in three places would mean it burns three times as fast as from one.

A single fuse, however, only has two ends, not three. You probably spotted that ...

But that's a detail we can get around. Cut the fuse into two pieces, and light both ends of the first piece, but only one end of the second. The fuse is now burning in three places, which is just what we wanted.

We need to make sure that the fuse is always burning in three places, so as soon as one of the pieces of the fuse burns out completely, cut the other piece in two, and light these two smaller pieces so that one of the pieces has two ends lit, and the other only one. Carry on this process until the remaining piece is too small to cut. Since the fuse has been burning continuously in three places until it is on the verge of burning out, it will have been lit for almost 20 minutes.

⑥⓪ THE BIASED COIN

This question was first asked – and solved – by John von Neumann, the Hungarian-born mathematical genius who made significant contributions to pretty much every scientific field he touched – as well as inventing some new fields himself.

A biased coin will not land on heads or tails with a 50/50 probability. Yet a biased coin flipped twice has an equal chance of landing on heads and then tails as it does of landing on tails and then heads. (More formally, if the probability of heads is a, and the probability of tails is b, the probability of heads then tails is $a \times b$, and the probability of tails then heads is $b \times a$, which is equal to $a \times b$.) So, to simulate a fair coin with your biased one, call either 'heads then tails (HT)' or 'tails then heads (TH)' and flip the coin twice. It will land either HT, TH, HH or TT. In the two latter cases, when the coin lands twice on the same side, ignore the result and flip twice again. Stop if you get an HT or a TH, but flip again with an HH or a TT. Carry on in this way until you get an HT or a TH. The chance of getting HT or TH is 50/50, which simulates the result of a fair coin.

⑥① DIVIDE THE FLOUR

Weighing 1: Balance the kilo of flour between the two pans, so that each pan contains 500g.

Weighing 2: Put one of these 500g piles of flour aside, and balance the remaining 500g piles between the two pans, giving 250g in each pan.

Weighing 3: Place one of these 250g piles aside. From the other, keep removing flour until what's left balances the 50g of the two weights. This leaves a 200g pile of flour. Gather the rest of the flour together, which will be 800g.

⑥② BACHET'S WEIGHT PROBLEM

We know that the following set of six weights will measure every number value from 1 to 63kg when put on only one pan:

1, 2, 4, 8, 16, 32

We're looking to find a smaller number of weights that measure from 1 to 40kg when we can put them on either pan. Let's start by measuring objects from 1kg upward with the minimum possible number of weights. We'll only introduce a new weight when we have to, and it will be the largest possible at each stage.

Let's call the two pans on a scale A and B.

To balance a 1kg object on A, we need a 1kg weight on B. So our set of kg weights so far is: 1.

To balance a 2kg object on A, we could use a 2kg weight on B. But there is another way that allows us to introduce a higher new weight. Since we already have a 1kg weight in our set, we could have the 2kg object *and* the 1kg weight on A, and balance them with a 3kg weight on B.

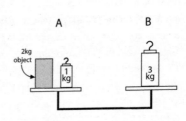

There are no other ways to balance 2kg with two different kg weights, so our set of kg weights is now: 1, 3.

With 1kg and 3kg weights we can measure objects up to 4kg. What is the highest new weight that permits us to weigh a 5kg object on A using both pans?

If we proceed as above and put the 5kg object on A with all our weights so far, 1kg + 3kg = 4kg, then we require a 9kg weight on B to balance the scales.

Our set of kg weights is now: 1, 3, 9.

With 1kg, 3kg and 9kg weights we can measure objects up to 13kg. What is the highest new weight that permits us to weigh a 14kg object on A using both pans?

Using the logic above, it will be 14kg + 13kg = 27kg.

Our set of kg weights is now: 1, 3, 9, 27.

And this will do us all the way up to 40kg. By using both sides of the pan we have reduced the set of weights from six to four.

You may have noticed a pattern: when weights can be placed on one pan, the numbers are the doubling sequence, where each new term is double the previous one. When weights can be placed on both pans, each term is three times the previous one. In the same way that the doubling sequence relates to binary numbers, the tripling sequence relates to base three, or *ternary* numbers, which are numbers that use only 0, 1 and 2.

For example, 1020 in ternary means no units, a 2 in the threes column, no nines and a 1 in the twenty-sevens column. And so, since 6 + 27 = 33, the number 1020 in ternary is equal to 33 in the decimal system.

⑥③ THE COUNTERFEIT COIN

Let's number the coins from 1 to 12.

Our first weighing is 1, 2, 3, 4 versus 5, 6, 7, 8.

If the pans balance this means that the counterfeit coin is among those we left out: 9, 10, 11 or 12.

Take three of these coins and weigh them against any three from the first weighing, which we know are all good coins. Let's say 1, 2, 3 versus 9, 10, 11.

If the pans balance we know that the counterfeit is 12, and for our third and final weighing we can balance it against any other coin to deduce if it is heavier or lighter.

If the pans don't balance, then the counterfeit is either 9, 10 or 11, and whether it is lighter or heavier depends on whether the pan with 9, 10 or 11 rose or fell. Now take any two of 9, 10, 11 and weigh them against each other, leaving the third out – as we did in the scenario with only nine coins mentioned in Bachet's Weight Problem. If the pans balance, the coin left out is the counterfeit. If we know the counterfeit is lighter, then if the pans don't balance, the coin in the pan that rises is the fake. If we know the counterfeit is heavier, then if the pans don't balance, the coin in the pan that falls is the fake.

The solution gets a bit more complicated if the original weighing of 1, 2, 3, 4 versus 5, 6, 7, 8 does not balance.

Imagine that 1, 2, 3, 4 falls against 5, 6, 7, 8.

We know that 9, 10, 11 and 12 are all good coins.

For the second weighing, take one of the good coins – say, 9 – and put it in a pan with two of the fallen coins – say 1 and 2. Put the other two from the fallen pan, 3 and 4, in with one from the risen pan, say 5. The coins numbered 6, 7 and 8 are excluded from this weighing.

The three possible outcomes are:

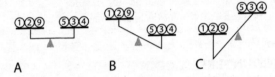

A **B** **C**

[A] The pans balance. So the counterfeit is one of 6, 7 or 8. For the third weighing, put 6 against 7. A balanced scale means the counterfeit is 8, which is lighter, since the pan with 8 rose in the first weighing. If 6 rises, it is the counterfeit, and if it falls 7 is the culprit.

[B] The left pan rises. The counterfeit must be among 1, 2, 3, 4 and 5, so we can eliminate 6, 7 and 8.

If one of 1, 2, 3, 4 is the counterfeit, then it must be heavier than normal, because the pan with 1, 2, 3, 4 fell in the first weighing. So the counterfeit must be either 3 or 4. Weigh them against each other in the third weighing.

[C] The right pan rises. As above in [B], the counterfeit must be one of these five coins, so we can eliminate 6, 7 and 8. And again, if one of 1, 2, 3, 4 is the counterfeit, then it must be heavier than normal, because the pan with 1, 2, 3, 4 fell in the first weighing. So the counterfeit may be 1 or 2.

But there is one possibility left. Since 5, 6, 7, 8 rose in the first weighing, the 5 coin could be the counterfeit and *lighter*.

So, in the final weighing, weigh 1 versus 2. The one that falls is the counterfeit, but if they balance the culprit is 5.

If in the first weighing the pan with 1, 2, 3, 4 rises against 5, 6, 7, 8, the process is the same as above, but swap 1, 2, 3, 4 for 5, 6, 7, 8.

64 THE FAKE STACK

One, of course!

On the pan place a single coin from the first stack, two from the second, three from the third, four from the fourth and so on until you've placed all ten from the final stack. There will be $1 + 2 + 3 + 4 \ldots + 10 = 55$ pound coins on the pan.

You know the weight of a single coin, so you know the weight of 55 coins. The difference in grams between the weight displayed on the scales and the weight of 55 coins is the number of the fake stack. If the difference is 1g, there is a single fake coin in the pan, meaning the fake stack is the first one; if the difference is 2g, there are two fake coins in the pan, meaning the fake stack is the second one; and so on.

65 FROM LE HAVRE TO NEW YORK

When I first heard this puzzle I instantly thought that the answer was 7, which was surely also the response of the illustrious French mathematicians who first encountered this puzzle from Monsieur Lucas.

The crossing takes seven days. So, the seven liners you will pass will be the ones leaving New York today, tomorrow, and so on until the one that leaves a day before you arrive.

Wrong! What about the liners that left New York in the past week? They are currently at sea and you will pass all of them on your journey too. The correct answer is that as you leave Le Havre you pass one in the port (that's the one arriving at noon, having left New York a week ago); you will pass 13 at sea; and you will meet a final one that is departing as you reach New York at noon in a week's time.

The diagram opposite explains it:

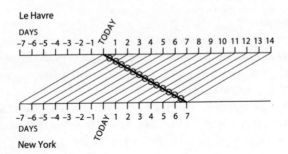

If the liners are all travelling at constant speed, you will pass a liner every 12 hours, once at noon (Le Havre time) and once at midnight.

66 THE ROUND TRIP

Consider the situation when the wind is blowing in the direction of the outward trip, and against the plane for the return.

Instinctively, it feels that the push of the tailwind and the drag of the headwind should cancel each other out, because what the wind giveth in one direction, in the other it taketh away. If the wind speed is W, then the speed on the way out increases by W and the speed on the way back reduces by W.

Yet what is relevant to this question is not flying speed but flying *time*. The amount of time gained by flying faster is not as great as the time lost by flying slower, because the plane will fly slower for longer.

We can plug in some numbers to see:

A plane travelling at 500mph travels 500 miles in an hour.

If the plane travels the same distance 100mph faster it gets there 10 minutes earlier. (Time = distance/speed. So, if the distance is 500 miles, and the speed is 600mph, which is 10 miles per minute, the journey in minutes is 500/10 = 50.)

If the plane travels the distance 100mph slower it gets there 15 minutes later.

(If the distance is 500 miles and the speed 400mph, the journey in hours is 500/400 = 1.25, which is an hour and a quarter.)

So a 500 mile trip there and back with a 100mph wind would take five minutes longer overall than if there was no wind at all.

But even without getting our hands dirty with calculations, imagine what would happen if the wind speed was equal to the speed of the plane. On the outward journey the plane would get there in half the time. But for the return journey the speed would be zero, so the plane would never leave the tarmac!

Okay, so this is an extreme case, but it reveals the pattern. By increasing speed by a fixed amount you are never going to be able to gain more than an hour, but by decreasing speed by a fixed amount you may end up losing a lifetime. So, summing up the above, a round trip when the wind is in the outgoing direction of travel will take longer than a round trip with no wind.

But what about crosswinds? A crosswind can be separated into a wind in the direction of travel (or against it), and a wind perpendicular to the direction of travel. Since we know that the former increases round trip time, what about the latter? In order to fly in a straight line from A to B when there is a perpendicular crosswind, the pilot needs to fly at an angle into the wind, which means that some of the velocity will be spent fighting the wind – rather than being used exclusively to get from A to B. So the trip there and the trip back will also take longer.

Best-case scenario for a round trip: no wind.

(67) THE MILEAGE PROBLEM

Your first realisation should be that the first four digits of the meters will only be the same again after the trip meter resets the clock to 000.0 when a thousand miles have been driven.

At this stage 876.6 miles (which we get by calculating 1000 – 123.4) will have been driven and the dashboard will look like this:

After another 130 miles the first two digits
are now the same:

And when another 3.5 miles have been
driven we have our answer:

The first four digits are again the same – and the car has driven 1010.1 miles.

68 THE OVERTAKE

These puzzles are not difficult – but precisely because they are easy our lazy
brains will not think them through properly.

[1] When you overtake the person in second place, you are now in second place.
[2] It is impossible to overtake the person in last place – since if he has someone
behind him he would not be last.

69 THE RUNNING STYLES

This is another fantastically counter-intuitive puzzle because it seems impossible
for Daphne, who runs every mile in 8 minutes and one second, to beat Constance,
who runs every mile in 8 minutes flat.

But, of course, she can – otherwise there would be no puzzle. The runner who
is slower over every mile can surprisingly devise a way to be faster over 26.2 miles.

Let's first think about how Daphne runs the race. She does not run at a
constant speed, but even so she covers every mile interval in the same time. How
does she do this?

Look at the graph below, which tracks a runner's speed over the distance of the course. It shows a racing strategy of running the first a of each mile at a high constant speed, and the remaining section of the mile at a low constant speed. If this strategy is repeated every mile all the way through the race, the runner is not running at a constant speed overall, but every mile interval will be run in the same time. This is because whichever mile you take, a of that mile is run at high speed and the remainder of the mile at low speed.

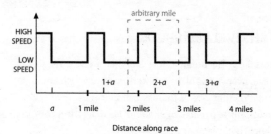

Distance along race

If Daphne runs a marathon using this strategy, then let's adapt it to best suit her needs. The marathon is 26.2 miles. We know – since she loses a second every full mile to Constance – that she will be 26 seconds behind at the 26-mile mark. So Daphne needs to make up 26 seconds on the remaining 0.2 miles.

Let Daphne's strategy be to run her high speed for the first 0.2 of each mile and her low speed for the rest. We now need to switch to talking about *time* rather than *distance*. Let Daphne run the first 0.2 miles in x seconds, and the remaining 0.8 miles in y seconds. She runs every mile in $x + y$ seconds.

The question states that Daphne runs every mile in 8 minutes 1 second, which is 481 seconds. This gives us the equation:

[1] $x + y = 481$

Constance runs every mile in 8 minutes, so she runs the entire marathon in $26.2 \times 8 \times 60$ seconds, which is 12,576 seconds.

Let's now assume that Daphne wins the marathon by a single second, completing it in 12,575 seconds. Since she runs the first 0.2 miles of her mile cycle 27 times, and the 0.8 mile section only 26 times, we get the following equation:

[2] $27x + 26y = 12,575$

We can solve these simultaneous equations without too much trouble. From [1] $y = 481 - x$. Substitute this value of y in [2]. The numbers thankfully cancel out nicely to leave $x = 69$ seconds and $y = 412$ seconds. So, if Daphne sprints the first 0.2 of each mile in 69 seconds, and takes 412 seconds for the rest of the mile, she will lose every single mile versus mile comparison, but pip Constance to the final post by a second.

(70) THE SHRIVELLED SPUDS

The answer is 50kg. The potatoes lose half their weight! The puzzle is interesting because the answer seems like a counter-intuitively high amount.

The calculation that gets us there is a simple one.

Water makes up 99 per cent of potatoes. Let's call the remaining 1 per cent the 'potato essence'.

When the weight of the potatoes is 100kg, this is made up of 1kg of potato essence and 99kg of water. The ratio of 'potato essence' to water is 1 to 99. After evaporation only 98 per cent of the potatoes is water, so the ratio of 'potato essence' to water becomes 2 to 98, which is the same as the ratio 1 to 49. Since the amount of 'potato essence' remains at 1kg, the amount of water must reduce to 49kg, so the total weight of the potatoes is 1kg + 49kg = 50kg.

The lesson here is that percentages are notoriously tricky to understand. The question deliberately sets us up to think that the change is tiny, from 99 to 98 per cent, a $\frac{1}{99}$ reduction. But the relevant change is from 1 to 2 per cent, a factor of two.

Often when it comes to percentages it is easier to think about actual things. Imagine restating the question as follows. A room has one woman and 99 men in it. A certain number of men leave, with the result that the woman goes from making up 1 per cent of the people in the room to making up 2 per cent. How many people remain in the room? The answer is half as many as there were before – just 50 people.

(71) THE WAGE WAGER

You might be able to second-guess this question. Plan B looks so much better, so of course it has to be less good.

Plan A
The starting salary is £10,000 annually. Every six months the six-monthly salary grows by £500. So, in the first six months you earn £5,000, at which point your salary rises by £500 for the next six months, at which point it rises again by £500, and so on. Here's what happens in the first two years.

	6 months		6 months		Total salary
Year 1	£5,000	+	£5,500	=	£10,500
Year 2	£6,000	+	£6,500	=	£12,500

Plan B
The starting salary is again £10,000, but the rise is only given at the end of the year. So, the salary for the first year is £10,000 and Plan B already lags behind Plan A. The mega-raise of £2,000 is added to the second year's salary, so at the end of the second year the salary is £12,000.

	Total salary
Year 1	£10,000
Year 2	£12,000

However, A remains in the lead, as it will in all subsequent years. The cumulative effect of smaller, more frequent raises beats the annual big one.

(72) A STICKY PROBLEM

If the point where the stick is to be cut is chosen at random, then any point on the stick has an equal chance of being chosen.

Half the time, therefore, the point will fall on the left half of the stick, and half the time on the right half. (We can ignore what happens when it falls exactly in the middle, because this position splits the stick into two identical sizes, so there is no smaller part.)

Now consider what happens when the cut is made on the left half of the stick. The smaller piece of the stick will be on the left, and its length will be between zero and half the length of the stick. In fact, since all points on the half stick have an equal chance of being chosen as the cutting point, the length of the smaller piece of the stick will be, on average, half the length of the half stick, which is a quarter of the length of the whole stick. The same argument can be used when the point falls on the right side of the stick. So the length of the smallest piece is a quarter of the length of the stick.

(73) THE HANDSHAKES

At the dinner party there are ten people in total: Edward, Lucy and four couples. So, the maximum number of people any single person can shake hands with is nine – everyone but themselves.

However, we are told that no one shakes hands with someone they know. We can assume that everyone knows their spouse, so the maximum number of people any single person can shake hands with is eight.

Edward received nine different answers, which therefore must have been 0, 1, 2, 3, 4, 5, 6, 7 and 8.

Consider the person who replied 8. They shook hands with everyone except their spouse. So everyone except their spouse shook at least one person's hand. The person who answered 0, therefore, must be in a couple with the person who replied 8.

Likewise, consider the person who replied 7. They shook hands with everyone except their spouse and the person who answered 0. So everyone else shook hands with at least two people – the person who replied 8 and the person who replied 7. So the spouse of the person who answered 7 must have answered 1.

Continuing this process we can deduce that the third couple consists of the 6 and the 2 and the fourth couple the 5 and the 3. The only person left, who must be Lucy, shook 4 people's hands.

(74) THE HANDSHAKES AND THE KISSES

Let's start with the handshakes. All men shake hands with other men. So, if there is a single male guest, there will be a single handshake, between him and Edward. If there are two male guests, there will be three handshakes – between the male guests, and one between Edward and each guest. And if there are three male guests there will be six handshakes. I'll leave you to work that out.

So, we know that three of the guests are men.

Women kiss everyone but their husbands. We know that three guests are men, so already we have three kisses accounted for – between Lucy and each of these men.

There are a total of twelve kisses, so we need new female guests to account for the other nine kisses. Let's introduce a single new female guest, Anna. If she is single she will kiss Edward, Lucy and the three men, making up 5 kisses. If she is married to one of the men, she will only kiss 4 people. We are still way off nine kisses, so let's introduce a second woman, Beatrice.

If Beatrice kisses everyone, she will kiss 6 people, but if she is married she will kiss only 5. Can the combined kisses of Anna and Beatrice add up to the nine we require? Yes, if both women are accompanied, since this gives 4 and 5 kisses. We're done. Five guests – two couples and a single man – attended dinner.

(75) THE LOST TICKET

This puzzle is less complicated than it seems. The solution requires no calculations or equations. The difficulty is in deciding how you approach it, but once you find the method it is elegant and straightforward.

The question tells the story of 100 people taking their seats in a theatre. Let's tell that story person by person.

So we don't get confused, let A be the first person in the queue, and Z the last. So we can rephrase the question as: What are the chances of Z sitting in his seat if A sits on a seat at random?

I've ordered the seats in a line in the diagram below, and marked A's and Z's designated seats, which I'll call seat A and seat Z:

Let's think what happens if A sits either in seat A or in seat Z. If she sits in seat A, everyone else will also sit in their correct seats, including Z. (This is what would happen if she hadn't lost her ticket).

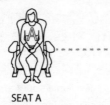

SEAT A   SEAT Z

A sits in seat A

And if she sits in seat Z, obviously Z will *not* sit in his correct seat, since A will be in it. In this case Z sits in seat A.

SEAT A   SEAT Z

A sits in seat Z

The question states that A chooses her seat randomly, so she has an equal chance of sitting in seat A or seat Z. So, if we restrict ourselves to just these two seats, then there is a 50/50 chance that Z will get to sit in his seat.

Now consider what happens if A sits on any other seat. Let's say that this is the seat of N, who is nth in the queue.

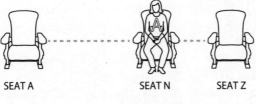

SEAT A   SEAT N   SEAT Z

A sits in seat N

If A sits in seat N, all the people in the queue up to N will sit in their designated seats. The first person not to sit in the correct seat is N, who cannot take his seat because A is in it. So N chooses one of the remaining seats at random.

The seats that are left for N to sit down in are seat A and the seats of the people in the queue after him, which includes seat Z.

So, N sits either in A (which means Z will eventually sit in the right seat) or in Z (which means Z will not sit in the right seat). Or N sits in M's seat, where M is an audience member after N but before Z.

If we restrict ourselves to thinking about whether N sits in A or Z, the chances are equal, so there is a 50 per cent chance that Z will get to sit in his seat. But what happens if N sits in seat M?

When it is M's turn to take a seat, she will face the same situation that N did: an equal chance to sit in either A or Z, or the chance to sit in the seat of a theatregoer who has yet to enter the auditorium. If M sits in the seat of a theatregoer yet to arrive, the same scenario repeats itself.

At every step of the way, the chance of any audience member choosing either seat A or seat Z is equal. And if that audience member chooses someone else's seat, then we are just delaying the decision to choose between A and Z to a subsequent theatregoer. Ultimately we will run out of audience members, and *someone* has to sit in either A or Z.

Any theatregoer who is forced to choose a seat randomly always has an equal choice between seat A and seat Z. And since choosing A means that Z sits in his correct seat, and choosing Z means that he doesn't, the chance of Z sitting in his own seat is 50 per cent.

Ten tasty teasers

Are you a genius at geography?

1) Rome.

2) Maine, because the northern Atlantic coast of the US stretches more towards the east than you might think.

3) Glasgow, Plymouth, Edinburgh, Liverpool, Manchester, since Scotland tilts towards the west.

4) Paris, Seattle, Halifax, Algiers, Tokyo.

5) Easter Island, Perth, Cape Town, Buenos Aires, Montevideo.

6) Germany. Its nine neighbours are, in clockwise order, Denmark, Poland, Czech Republic, Austria, Switzerland, France, Luxembourg, Belgium, Netherlands.

7) From smallest to largest: Falkland Islands, Shetland Islands, Isle of Man, Jersey, Isle of Wight.

8) Canada.

9) China has a single time zone, remarkably, even though its east–west distance is about 3,000 miles, just under twice the distance from London to Moscow.

10) Aconagua, 6,962m; Mount McKinley, 6,194m; Mount Kilimanjaro, 5,892m; Mount Elbrus, 5,642m.

Your Aid I Want, Nine Trees to Plant

PROBLEMS WITH PROPS

76 THE SIX COINS

To start, arrange the coins in a parallelogram pattern, as shown below. Each move is marked with an arrow.

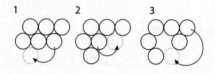

77 TRIANGLE TO LINE

(78) THE WATER PUZZLE

(79) THE FIVE PENNIES

Dudeney's solution was to lie one penny flat on the ground, with two others flat on top. The final two coins can be positioned such that they touch each other at the top and the other three at or near the bottom. It is very fiddly to do, but it works! In *The Tokyo Puzzles*, Kobon Fujimura said that one of his readers sent in a second solution, with only a single coin on its edge.

(80) PLANTING TEN TREES

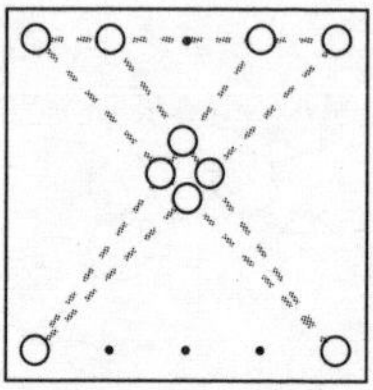

The puzzle starts with two lines of five coins, and the aim is to make five lines of four.

You can change the positions of only four coins. It makes sense to take a single coin from one of the lines of five, which gives the first line of four, and the remaining three coins from the other line of five.

In the illustration above right you can see that four new lines of four can be created by joining each coin on the first line of four to one of the two coins opposite. In this solution I moved the middle coin in the top row and the three middle coins in the bottom row. In fact, I could have moved any coin from the top (or bottom), and any three from the bottom (or top). Here are two more solutions with different coins chosen:

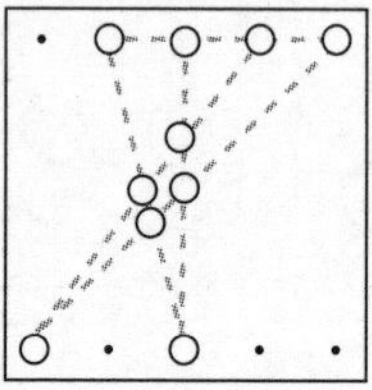

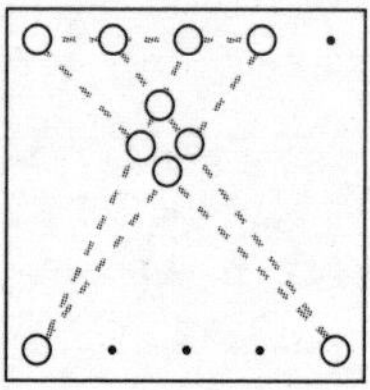

So, how many solutions are there in total? There are five ways to choose a single coin from a row of five, and there are 10 ways to choose three coins from a row of five, so there are $5 \times 10 = 50$ ways to choose a single coin from the top line and three from the bottom. This gives us 50 solutions. To this we can add another 50 solutions when we choose a single coin from the bottom and three from the top, giving us 100 solutions.

I'll accept this answer. However, extra marks for realising that each of these 100 solutions can be made in 24 ways, since the four moved coins can be placed in 24 possible ways. For example, look at the first solution above, where the moved coins are in a diamond formation. If these coins are A, B, C and D, and we begin with, say, the top position and move clockwise, then one solution is when they are arranged ABCD, another when they are arranged ABDC, another ACBD, and so on for the 24 permutations of A, B C and D.

The total number of solutions is therefore 100 × 24= 2,400.

And here is Dudeney's gallery of arboreal arrangements with ten trees in five lines of four.

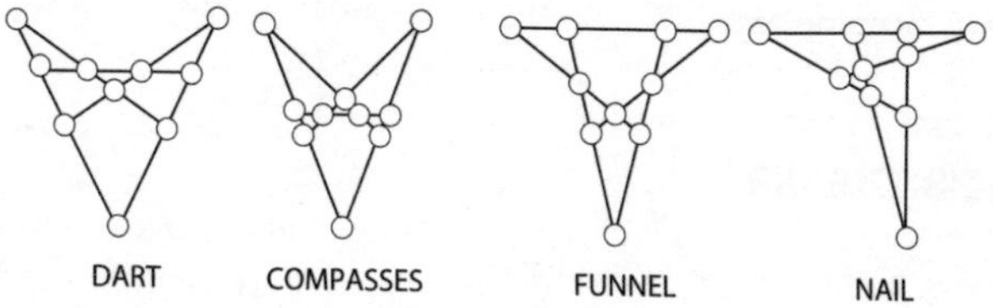
DART
COMPASSES
FUNNEL
NAIL

(81) THE SPACE RACE

The player who goes first will always win, if he or she applies the following strategy.

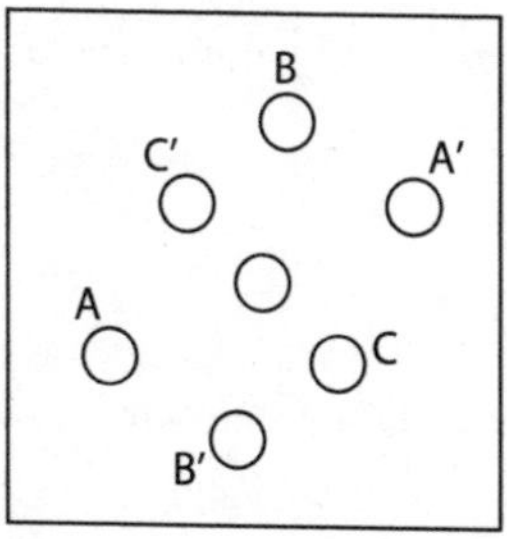
B
C′
A′
A
C
B′

Player 1 places the first coin dead centre in the table, and in all subsequent moves he or she simply places a coin in the opposite position to where Player 2 has just played. So, in the diagram on the right, if Player 2 places a coin at A, Player 1 then places at A'. And likewise if Player 2 places a coin at B or C, Player 1 copies the move by placing a coin at B' or C'.

Since the table starts empty, wherever Player 2 places a coin it will always be possible for Player 1 to put a coin in the opposite position. So Player 1 cannot lose, and eventually Player 2 will find himself with no free table left.

If you want to play this game with cigars, you must stand the very first cigar on its end. You cannot place the first cigar on its side, because the ends of cigars are not the same – one is flat and the other is tapered. (Now you are thanking me for replacing cigars with coins. The rotational asymmetry of cigars is not knowledge one can readily assume these days, even in London clubs.)

If Player 1 places the cigar on its side in the middle, as shown right, but Player 2 places a cigar at D, very close to the tapered edge, then Player 1 will be unable to place a cigar opposite at D' without it touching the central cigar. Coins do not have this problem.

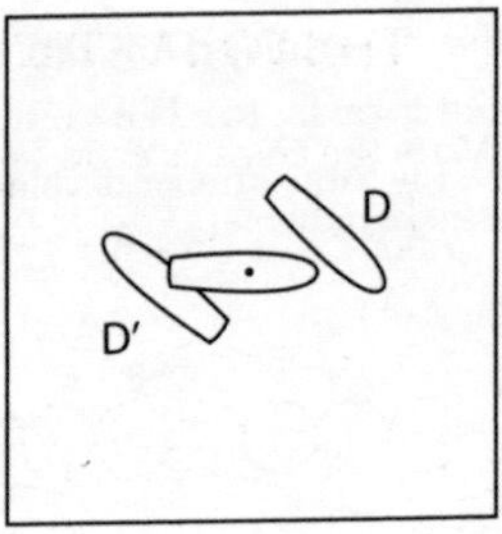

(82) TAIT'S TEASER

Tait's original question is solved as follows:

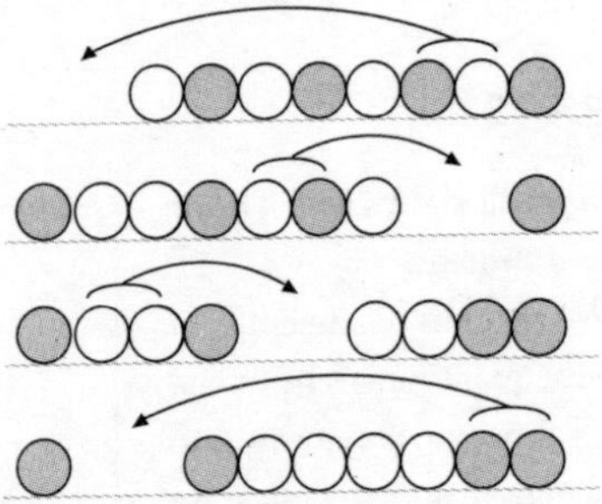

And the one with only five coins:

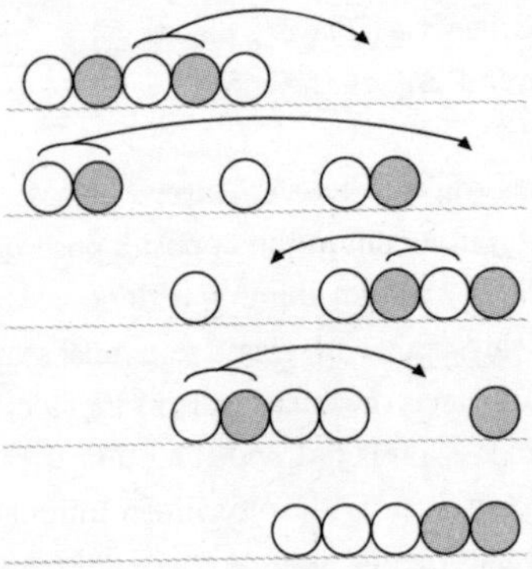

(83) THE FOUR STACKS

Move the coins as show below, following the numbers. The double coins are the stacks of two.

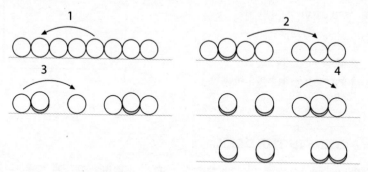

(84) FROGS AND TOADS

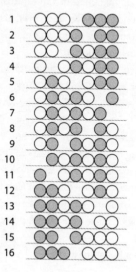

85 TRIANGLE SOLITAIRE

Remove the coin on position 2 as we did before. Curiously, the first and last moves are the same as for the six-move solution. The clever bit is not to complete all jumps on the third move.

1. Move 7 to 2.
2. Move 1 to 4.
3. Move 9 to 7, then to 2.
4. Move 6 to 4, then to 1, then to 6.
5. Move 10 to 3.

86 COINS IN THE DARK

If the audience member tells you that there are x heads, then you will be able to separate the coins into two groups with the same number of heads by selecting any x coins and flipping them over.

So, for example, if you are told there are three heads, your strategy is to choose any three coins to be in one group, flip them over, and they will have the same number of heads as the remaining coins. Your choice will work however many heads are in the three coins you selected.

Likewise, if you are told there are five heads, choose any five coins to be in one group, flip them over, and they will have the same number of heads as the remaining coins. Again, your choice will work however many heads are in the five coins you selected.

Note that it is impossible to know how many heads are in the group you turn over, and in the remaining coins, but then you didn't promise that. All you can say for certain is that the number of heads in each of the groups is the same. It is a remarkably simple solution for something that feels like magic.

Try it a few times by having, say, three coins as heads, and then choosing lots of different combinations of three coins and flipping them. You will begin to understand why it works.

The proof, however, requires some algebra.

Let's say you are told there are x heads in the ten coins. Choose any x coins, and call them Group A. If all of these coins are heads, then the remaining coins, Group B, are all tails. So if you flip all of Group A, the coins will all be tails and both groups will have the same number of heads, which is zero.

But let's say there are no heads in Group A. This means that the x heads are all in Group B. If we flip every coin in Group A, it will then contain x heads, and again both Groups have the same number of heads, which is x.

Now let's say that some of the coins in Group A are heads, and some tails. If there are y tails in Group A, then there are $(x - y)$ heads in Group A, which means there must be y heads in Group B. So by flipping all the coins in Group A it will contain $(x - y)$ tails and y heads, equalling the number of heads in Group B.

The trick works for any number of coins, not just ten. If you know the total number of heads, you can separate the coins into two groups that contain an equal number of heads by selecting a number of coins that equals the number of heads and flipping them.

(87) THE ONE HUNDRED COINS

This solution relies on the fact that one hundred is an even number.

Number the coins from 1 to 100. If Penny goes first she can guarantee that she will collect all the odd coins, or all the even coins. For example, if she wants all the odd coins, she will start by taking coin 1. Bob will choose either 2 or 100, but whichever one he removes he will leave an odd numbered coin for Penny's next choice. When she takes this odd coin, she leaves Bob with only even coins at both ends. He is again forced to take an even coin, and this split between Penny taking odds and Bob taking evens will continue until there are no coins left. Similarly, if Penny wants all the even numbers, she will take 100 first. Bob will choose either 1 or 99, leaving an even coin for Penny's next move, and so on.

Penny's strategy, therefore, is to add up the values of all the odd coins, and all the values of the even coins, and to choose odd or even based on whichever sum is the highest total value. If the value of the odd coins is different from the value of the even coins she is guaranteed to win. If the value of the odd coins is equal to the value of the even coins, then by choosing either odd or even she will collect the same money as Bob. So she will always make at least as much money as he does.

The fascinating, and seemingly paradoxical, consequence of this game is that if we add an extra coin, so there are 101 in total, the advantage will probably now fall to Bob, even though he will be collecting fewer coins! Once Penny has chosen her first coin, leaving only 100 in the row, Bob will proceed as Penny did above. He will tally the odd ones, and the even ones, and choose to collect whichever sum is greatest. The only time he will lose is if the difference between the odd and even sums is less than the value of Penny's first coin.

It is striking to think that the oddness or evenness of the row of coins – rather than the values of the coins or the total number of them in the row – makes such a pivotal difference to who will win.

88 FREE THE COIN

Using a second match, light the head of the match and quickly blow it out. It will stick to the glass on the right, enabling you to remove the glass on the left and take the coin.

89 PRUNING TRIANGLES

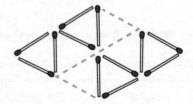

90 TRIANGLE, AND TRIANGLE AGAIN

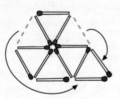
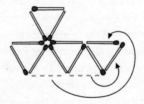

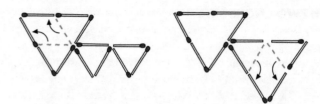

91 GROWING TRIANGLES

[1]

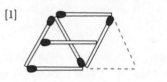

[2]

92 A TOUCHING PROBLEM

93 POINT TO POINT

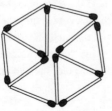

(94) THE TWO ENCLOSURES

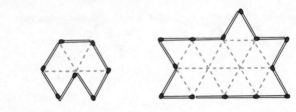

(95) FOLDING STAMPS

To get the order 1-5-6-4-8-7-3-2, proceed as follows:

Step 1. Fold the stamps so that the back of 6 and the back of 7 are touching. Keep them together by pinching the front of 6 and 7 with your forefinger and thumb.

Step 2. With your other hand, fold the front of 4 onto the front of 8. Keep the 4 and the 8 pinched together with forefinger and thumb.

Step 3. Bend the 4 and the 8 and slide them in between the 6 and the 7. Now the stamps 6, 4, 8 and 7 are in the right order.

Step 4. Straighten the section that has 1, 2, 5 and 6. Fold the front of the 5 onto the front of the 6, and the problem is solved.

To get the order 1-3-7-5-6-8-4-2, wrote Dudeney, 'is more difficult and might well have been overlooked, if one had not been convinced that, according to [a law I discovered], it must be possible.'

Step 1. Fold the block in half along the central horizontal line so the fronts of 1, 2, 3 and 4 are visible on the front, and 5, 6, 7 and 8 are visible on the back.

Step 2. Fold the face of 5 onto the face of 6.

Step 3. With one hand, hold the stamps so that the thumb is on 1 and the forefinger on 2. With the other hand hold the other end of the block, which has the 8 on the front and the 4 on the back. The tricky part is to now thread the 8/4 end between the 1 and the 5, and then to continue it round so the 8/4 end slips in between the 6 and the 2, leaving only the 3 and the 7 between the 1 and the 5. The job is done!

⑨⑥ THE FOUR STAMPS

Four stamps may be torn from the block in each of the following ways:

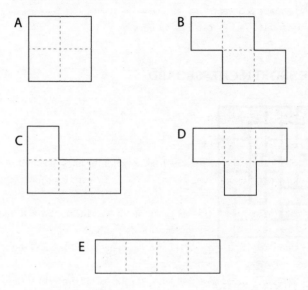

You need to be careful when tallying the different shapes to take care of all the possible orientations and rotations.

Shape A can be made in **6** ways.

Shape B can be made in 4 ways when shown as above, and 3 ways when rotated 90 degrees. If you flip it around a vertical axis so the Z becomes an S, there are another 4 ways, and 3 more when the S is rotated 90 degrees, making **14** ways in all.

C can be made 4 ways when shown as above, 3 ways when rotated 90 degrees, 4 ways when rotated 180 degrees, and another 3 ways when rotated 270 degrees, making 14 ways. If C is flipped around a vertical axis as shape B was, we get another 14 ways, making a total of **28**.

D can be made 4 ways when shown as above, 3 ways when rotated 90 degrees, 4 ways when rotated 180 degrees and 3 ways when rotated 270 degrees, giving a total of **14** ways.

E can only be made in **3** ways.

The grand total is 6 + 14 + 28 + 14 + 3 = **65** ways.

(97) THE BROKEN CHESSBOARD

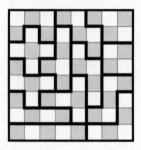

(98) FOLDING A CUBE

Once you fold the piece of paper after the fourth and the sixth squares, you will be able to make the cube easily.

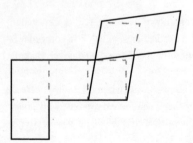

⑨⑨ THE IMPOSSIBLE BRAID

You can solve this puzzle in different ways. The quickest method requires two twists of the strands through themselves, which is how Scouts and Guides are taught to braid their woggles. I'm not going to show that method, however, since you would probably only find these twists through guessing. (If you are interested then search for the solution online.)

What I like about this puzzle, though, is that you can also make the braid using common sense. In fact, the easiest way to solve this puzzle is so straightforward that when you understand it, it is hardly a puzzle at all. All you needed to do was follow the instructions ...

I'm assuming we all know how to braid. You cross the leftmost strand over the central one, then the rightmost over the central one, then the leftmost, and so on. In the illustration below, 1 crosses 2, then 3 crosses 1, which is now in the centre. The next step would have 2 (on the left) crossing 3 (in the centre), and so on.

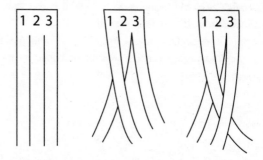

I told you that the strands in the braid cross six times. That was a clue. For a moment let's forget that the three strands join at the top and the bottom ends of the strip. Start at the top end and braid. Cross 1 over 2, then 3 over 1, and continue for four more crossings until you have done six in all. (It is quite fiddly, which is why I recommended you use plastic. Paper may rip.) If you pinch the sixth crossing between your thumb and your forefinger, you should have something that looks like the crazy mangle below.

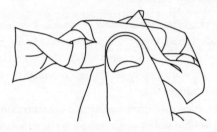

The plastic looks like this because each time we made a nice braid crossing at the top end of the strip, it was countered by an ugly twist on the other end. After six crossings, the side to the left of my thumb is the pattern we are aiming for in the solution – and the side to the right is a scrunch of plastic.

What now? Well, try to untangle the mess to the right of the finger and thumb using your other hand. If you thread the right end through itself a few times, the strands will untangle perfectly. Readjust the braid so it flows evenly down the strands. The impossible braid is possible after all.

This solution is not particularly elegant, but it works. Sometimes the answer to a question is the simplest response. The question said braid, so braid!

Ten tasty teasers

Are you smarter than a 13-year-old?

1) B

The statements are all contradictory, so at most one of them can be true. And if one of them is true it must be the second one, which is indeed true.

2) A

If the overlapping region is a triangle, two of its sides must be adjacent sides of a square, so one of its angles will be 90 degrees. This means that an equilateral triangle, in which all the angles are 60 degrees, is impossible. Here are how the other shapes can appear.

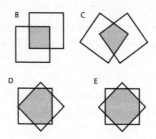

3) D

Consider the units digits on each side of the equations, as they will tell us which equations is correct. The units digit of $44^2 + 77^2$ is 5, since the units digit of 4^2 is 6 and the units digit of 7^2 is 9. The units digit of $55^2 + 66^2$, and $66^2 + 55^2$, is 1, and the units digit of $99^2 + 22^2$ is 5. So all those statements are false. Finally, we need to check that $88^2 + 33^2 = 7744 + 1089 = 8833$.

4) D

Clearly, there must be a minimum of two 'on' switches. Two 'on' switches and three 'off' switches may be set in only one way: 'off', 'on', 'off', 'on', 'off'. Three 'on' switches and two 'off' switches may be set in six different ways. Four 'on' switches and one 'off' switch may be set in five different ways. Finally, five 'on' switches can be set in just one way. So there are 13 ways in all.

5) E

Consider the thousands column. The letters represent different digits. Since S = 3, M must be either 0, 1 or 2. We can eliminate 0 and 1 since that would make MANY too small, so M = 2, and there must be a carry of 1 from the hundreds column. A = 9, since if it were any lower the sum wouldn't work. This means that U must be 0, since it cannot be 9, as 9 is already taken, and 0 is the only other option if we carry 1 over from the tens column. In the tens column, N must be an 8 with a carried 1, since it cannot be 9, as that's already been used. That leaves O + Y = 13. The only pairs of numbers that work for O and Y are 4 and 9 (or vice versa), 5 and 8 (or vice versa) and 6 and 7 (or vice versa). But 8 and 9 have been used up already, so it must be the last option, and $6 \times 7 = 7 \times 6 = 42$.

6) D

The only such occasions occur when the clock changes from 09 59 59 to 10 00 00; from 19 59 59 to 20 00 00 and from 23 59 59 to 00 00 00.

7) D

The first six positive cubes are 1, 8, 27, 64, 125 and 216. Clearly 64 cannot be the sum of three positive cubes as the sum of all the positive cubes smaller than 64 is $1 + 8 + 27 = 36$. Similarly, 125 cannot be the sum of three positive cubes as the largest sum of any three positive cubes smaller than 125 is $8 + 27 + 64 = 99$. However, $27 + 64 + 125 = 216$, so 216 is the smallest cube which is the sum of three positive cubes.

8) C

If the first three terms are –3, 0 and 2, the fourth term is –3 + 0 + 2 = –1. The fifth term is therefore 0 + 2 –1 = 1, and so on. The first thirteen terms of the sequence are –3, 0, 2, –1, 1, 2, 2, 5, 9, 16, 30, 55, 101, ...

9) C

Pages 1 to 9 inclusive require 9 digits; pages 10 to 99 inclusive require 180 digits. So 189 digits are required to number all the pages before the three-digit numbers start on page 100. This leaves 663 digits, which account for another 221 pages. So the book has 9 + 90 + 221 = 320 pages.

10) B

Imagine the cross to consist of three horizontal layers. The first layer contains only the cube which was glued to the top face of the original cube. The second layer contains the original cube plus four additional cubes glued to its side faces. The third layer contains only the cube which was glued to the bottom face of the original cube. When yellow cubes are now added, one cube will be glued to the top face of the blue cube on the first layer and four to its side faces. Eight yellow cubes will be glued to the blue cubes in the second layer. And the single blue cube in the third layer will have five yellow cubes glued to it, like the one in the first layer. So, overall, 18 yellow cubes are required.

The Number Games

PROBLEMS FOR PURISTS

⑩ MIRROR, MIRROR

The sums are the same! It's a result that seems really rather surprising until you consider the calculation column by column. Maybe say it aloud. The first column in the left-hand sum contains one nine, or 1×9; the first column in the right-hand sum contains nine ones, or 9×1. The second column in the left-hand sum contains two eights, or 2×8; the second column in the right-hand sum contains eight twos, or 8×2. And so on. Each column adds to the same number, so the total sums are the same.

⑩ NOUS LIKE GAUSS

If we were to list all the numbers on top of each other, as you do when you're writing out addition, the Gaussian insight is to realise that each column – the units column, the tens column, the hundreds column and the thousands column – contains the same digits, which are six 1s, six 2s, six 3s and six 4s, even though the order of the digits will be different in each column. The sum of each column is easy to calculate: it's $(6 \times 1) + (6 \times 2) + (6 \times 3) + (6 \times 4) = 6 + 12 + 18 + 24 = 60$. So the total sum is

```
   60
   60
   60
   60
 ‾‾‾‾‾
 66660
```

⑬ THAT'S SUM TABLE

You may have solved this in one of two ways. I'll call the first the Alcuin method – since it is most faithful to how he paired the numbers when summing from 1 to 100 – and the second the Gauss method.

1	2	3	4	5	6	7	8	9	10
2	3	4	5	6	7	8	9	10	11
3	4	5	6	7	8	9	10	11	12
4	5	6	7	8	9	10	11	12	13
5	6	7	8	9	10	11	12	13	14
6	7	8	9	10	11	12	13	14	15
7	8	9	10	11	12	13	14	15	16
8	9	10	11	12	13	14	15	16	17
9	10	11	12	13	14	15	16	17	18
10	11	12	13	14	15	16	17	18	19

Alcuin method. Pair the numbers diagonally from top left to bottom right. You'll see that $(1 + 19) = 20$, $(2 + 18) = 20$, $(3 + 17) = 20$, and so on until $(9 + 11) = 20$. There is one of the first pair, two of the second pair, three of the third pair, and so on. So the sum of these pairs is $20 + (2 \times 20) + (3 \times 20) + ... + (9 \times 20)$, or $(1 + 2 + 3 + ... + 9) \times 20$, which is $45 \times 20 = 900$. To this we add the ten 10s in the diagonal that we haven't yet counted. So the total is $900 + 100 = 1,000$.

Gauss method. The sum of the first row is equal to $(1 + 10) + (2 + 9) + ... + (5 + 6) = 5 \times 11 = 55$. The numbers in the second row are all +1 of the numbers in the first row, so the sum of the second row is equal to the sum of first row plus 10. The sum of the third row is the sum of the second row plus 10, which is the sum of the first row plus 20. The sum of the whole table is therefore:

$55 + (55 + 10) + (55 + 20) + ... + (55 + 90)$

Which is

$(10 \times 55) + (10 + 20 + 30 + ... + 90)$

or

$550 + 10(1 + 2 + 3 + ... + 9) = 550 + (10 \times 45) = 550 + 450 = 1,000$

⑩⁴ THE SQUARE DIGITS

Some numbers are ruled out of certain positions – such as 1, which cannot be in the multiplication, since when you multiply a digit by 1 you get the same digit and the digits here appear only once. But really the only way to solve this is by trial and error.

$9 - 5 = 4$
$\qquad \times$
$6 \div 3 = 2$
$\qquad =$
$1 + 7 = 8$

⑩⁵ THE GHOST EQUATIONS

$27 \times 3 = 81$
$6 \times 9 = 54$

⑯ RING MY NUMBER

Sums to 11:

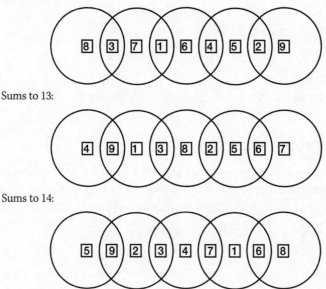

Sums to 13:

Sums to 14:

⑰ THE FOUR FOURS

There are often alternatives to the solutions shown here.

From 2 to 9:

$2 = \left(\frac{4}{4}\right) + \left(\frac{4}{4}\right)$

$3 = \frac{(4 + 4 + 4)}{4}$

$4 = 4 + (4 \times (4 - 4))$

$5 = \frac{((4 \times 4) + 4)}{4}$

$6 = \left(\frac{(4 + 4)}{4}\right) + 4$

$7 = 4 + 4 - \left(\frac{4}{4}\right)$

$8 = 4 + 4 + 4 - 4$

$9 = 4 + 4 + \left(\frac{4}{4}\right)$

294 | CAN YOU SOLVE MY PROBLEMS?

From 10 to 20:

$10 = \dfrac{44-4}{4}$

$11 = \dfrac{44}{(\sqrt{4}+\sqrt{4})}$

$12 = 4 \times (4 - (\frac{4}{4}))$

$13 = (\frac{44}{4}) + \sqrt{4}$

$14 = (4 \times 4) - 4 + \sqrt{4}$

$15 = (4 \times 4) - (\frac{4}{4})$

$16 = (4 \times 4) + 4 - 4$

$17 = (4 \times 4) + (\frac{4}{4})$

$18 = (4 \times 4) + 4 - \sqrt{4}$

$19 = \dfrac{((4+\sqrt{4})}{.4} + 4$

$20 = (4 + \frac{4}{4}) \times 4$

From 21 to 30:

$21 = 4! - 4 + (\frac{4}{4})$

$22 = 4 \times 4 + 4 + \sqrt{4}$

$23 = 4! - \sqrt{4} + \frac{4}{4}$

$24 = (4 \times 4) + 4 + 4$

$25 = 4! + \sqrt{4} + \frac{4}{4}$

$26 = 4! + \sqrt{(4+4-4)}$

$27 = 4! + 4 - (\frac{4}{4})$

$28 = ((4+4) \times 4) - 4$

$29 = 4! + 4 + (\frac{4}{4})$

$30 = 4! + 4 + 4 - \sqrt{4}$

From 31 to 40:

$31 = \dfrac{((4+\sqrt{4})! + 4!}{4!} = 31$

$32 = (4 \times 4) + (4 \times 4)$

$33 = 4! + 4 + (\sqrt{4}/.4)$

$34 = (4 \times 4 \times \sqrt{4}) + \sqrt{4}$

$35 = 4! + 44/4$

$36 = 44 - 4 - 4$

$37 = 4! + (\dfrac{(4! + \sqrt{4})}{\sqrt{4}})$

$38 = 44 - (4!/4)$

$39 = 4! + \dfrac{4!}{(4 \times .4)}$

$40 = 4! - 4 + 4! - 4$

From 41 to 50:

$41 = \left(\frac{(4! + \sqrt{4})}{.4}\right) - 4!$

$42 = 44 - 4 + \sqrt{4}$

$43 = 44 - (4/4)$

$44 = 44 + 4 - 4$

$45 = 44 + \left(\frac{4}{4}\right)$

$46 = 44 + 4 - \sqrt{4}$

$47 = 4! + 4! - \frac{4}{4}$

$48 = (4 + 4 + 4) \times 4$

$49 = 4! + 4! + \frac{4}{4}$

$50 = 44 + 4 + \sqrt{4}$

(With thanks to mathforum.org, whose table of answers I cribbed here.)

⑽ OUR COLUMBUS PROBLEM

$$
\begin{array}{r}
80.\dot{5} \ \left(\text{or } 80\frac{55}{99}\right) \\
.\dot{9}\dot{7} \ \left(\text{or } \frac{97}{99}\right) \\
+ \ .\dot{4}\dot{6} \ \left(\text{or } \frac{46}{99}\right) \\
\hline
82
\end{array}
$$

The fractions cancel out nicely since $\frac{55}{99} + \frac{97}{99} + \frac{46}{99} = \frac{198}{99} = 2$

⑼ THREES AND EIGHTS

$24 = \dfrac{8}{\left(3 - \left(\frac{8}{3}\right)\right)}$

⑽ CHILD'S PLAY

If a puzzle is introduced with the statement that children can solve it faster than adults, that's a clue that no understanding is required beyond recognising the simplest visual patterns. On seeing a list of numbers, adults will automatically start to think numerically. Yet for this puzzle the numbers are simply shapes with no numerical significance at all. Count the number of loops in each of the four-digit numbers, and the tally is the number on the right of the equals sign. The symbol 8 has two loops, 0 has one and 9 has one, so the number 8809 has six loops. The number of loops in 2581 is therefore 2.

⑪ FOLLOW THE ARROW 1

I hope you didn't overthink. The rule here is simple. For each number, multiply the two digits together:

$7 \times 7 = 49$ $4 \times 9 = 36$ $3 \times 6 = 18$

So the number that comes next is $1 \times 8 = 8$.

⑫ FOLLOW THE ARROW 2

The rule is that you square each digit in a number, and add the results. So:

$4^2 = 16$ $1^2 + 6^2 = 37$ $3^2 + 7^2 = 58$

And so on, revealing that the missing number is 20, because:

$4^2 + 2^2 = 20$, and $2^2 + 0^2 = 4$

I solved this by staring at the $4 \rightarrow 16$ long enough, before realising that squaring *had* to be involved. The next step was to work out how squaring got me from 16 to 37. And then all the dominoes fell.

⑬ FOLLOW THE ARROW 3

This one is *really* tricky if you haven't seen it before. There seems to be no arithmetical rule that fits.

But perhaps you noticed when mouthing the numbers to yourself that the words for the numbers are getting longer:

Ten
Nine
Sixty
Ninety
Seventy
Sixty-six

When it's written down, the pattern is clear. The first number in the list has three letters, the second has four letters, the third five and the others six, seven and eight. The numbers are listed in order of word length, with each new term increasing the count by one letter.

So, the next term has nine letters. But many numbers have nine letters! For example: *Forty-four. Fifty-five. Sixty-nine. Ninety-six.* Which one is it to be?

Back to the list for more thought.

The only three-letter numbers are one, two, six and ten.

The only four-letter numbers are four, five and nine.

Each number in the sequence is the *largest number* with that number of digits. We can verify the others on the list too.

The largest number with nine letters is ninety-six, so this is the answer.

Well, not quite. The number:

10000000000000000000000000000000000000000000000000000000000000000
0000000000000000000000000000000000000000, or 10^{100}, is a 'googol', and it has six letters. Add an extra zero and the number is *ten googol*, which has nine letters. That's the best answer.

And it explains why this question, allegedly, used to be a favourite interview question at Google.

⑪⑭ DICTIONARY CORNER

There are one quadrillion numbers in the dictionary. All numbers must begin with one of the words 'one', 'two', 'three', 'four', 'five', 'six', 'seven', 'eight', 'nine', 'ten', 'eleven', 'twelve', 'thirteen', 'fourteen', 'fifteen', 'sixteen', 'seventeen', 'eighteen', 'nineteen', 'twenty', 'thirty', 'forty', 'fifty', 'sixty', 'seventy', 'eighty' or 'ninety'.

The first entry must therefore be 8.

With similar reasoning, the last entry has to start with a 2, since 'two' is the furthest down the alphabet of the words listed above. The answer is not 2, however, since numbers that begin with two (but are not 2) will come after two in the dictionary. The options for the next word of our number are 'trillion', 'billion', 'million', 'thousand' and 'hundred'. Trillion is furthest down the alphabet, so the first two words of this number are 'two trillion'. The next word in this number must also be two. Thousand. Two. Hundred. Two. The answer is 2,000,000,002,202.

The first odd entry must start with an 8, but is obviously not eight since 8 is even. Of the options for the next possible word – again 'trillion', 'billion', 'million', 'thousand' and 'hundred' – the closest to the beginning of the alphabet is 'billion'. The next five letters have to be 'eight' – and the choices are 'eighteen', 'eighty', 'eight million', 'eight thousand', or 'eight hundred'. Eighteen wins. We carry on like this and get the following. Million. Eighteen. Thousand. Eight. Hundred. Eighty. At which point the number is 8,018,018,88X, where X is the final digit. The number is odd, so we must choose from 'one', 'three', 'five', 'seven' and 'nine'. Five wins.

The answer is 8,018,018,885.

The last odd entry follows the same process as above, to get 2,000,000,002,203.

BONUS PROBLEM: **SEND MORE MONEY**

We have got this far:

$$
\begin{array}{r}
9\ E\ N\ D \\
+\quad 1\ O\ R\ E \\
\hline
=1\ O\ N\ E\ Y
\end{array}
$$

If there is a carry in the thousands column, $1 + 9 + 1 = 1O$ (where O is upper case o), but this would make the upper case letter $O = 1$, which it cannot be since $M = 1$. So there is no carry in the thousands column, meaning $O = 0$. Which is helpful, since the confusion between zero and the letter o was getting distracting!

There must, however, be a carry in the hundreds column since otherwise $E + 0 = N$, meaning that $E = N$ which is not allowed since no two letters represent the same number. The sum now looks like this:

$$
\begin{array}{r}
\overset{1}{9}\ \overset{x}{E}\ N\ D \\
+\quad 1\ 0\ R\ E \\
\hline
=1\ 0\ N\ E\ Y
\end{array}
$$

I've also added an x in the space for the carry in the tens column. If there is a carry, $x = 1$ and if there isn't, $x = 0$. I put the x there because it means we can write the following three equations from the remaining columns:

Hundreds column: $E + 1 = N$

Tens column: $x + N + R = 10 + E$ (The ten represents the carry)

Units column: $D + E = Y + 10x$

If $x = 0$, then by substituting $E + 1$ for N in the second equation we get:

$E + 1 + R = 10 + E$, which simplifies to:

$R = 9$,

This result is impossible since $S = 9$. So, $x = 1$, which makes the three equations above:

$E + 1 = N$

$N + R = 9 + E$

$D + E = Y + 10$

By substituting E + 1 for N in the second equation we get E +1 + R = 9 + E, which simplifies to R = 8.

$$
\begin{array}{r}
9\ \overset{1}{E}\ \overset{1}{N}\ D \\
+\ \ \ 1\ 0\ 8\ E \\
\hline
=\ 1\ 0\ N\ E\ Y
\end{array}
$$

And we are left with:

E + 1 = N

D + E = Y + 10

The numbers 0 and 1 are already used up, so Y must be 2 or greater. So D + E > 12. Because 9 and 8 are already used, the only possible numbers for D and E are either 6 and 7 (or vice versa), or 5 and 7 (or vice versa).

Let's say they are 6 and 7. So, E is either 6 or 7. If E = 6, D = 7. But this leads us into contradiction since E + 1 = N, meaning that N = 7 also, and different letters do not represent the same digit. Alternatively, if E = 7, then the equation E + 1 = N tells us that N = 8, but 8 is already taken by R.

So, D and E are either 5 and 7 or 7 and 5.

But E cannot be 7, for the same reason as above, since this would mean that N = 8, which it cannot be. So D = 7, E = 5, Y = 2 and N = 6.

$$
\begin{array}{r}
9\ \overset{1}{5}\ \overset{1}{6}\ 7 \\
+\ \ \ 1\ 0\ 8\ 5 \\
\hline
=\ 1\ 0\ 6\ 5\ 2
\end{array}
$$

⑮ THE THREE WITCHES

Step 1. T must be 1, because if the sum of two six-digit numbers equals a seven-digit number, that number must begin with a 1. (We can ignore the contribution of the four digit number here, since it cannot tip the total sum into a seven-digit number beginning with a 2 or above. Since every letter is a different number, the highest possible value for DOUBLE + DOUBLE + TOIL is $987{,}543 + 987{,}543 + 6{,}824 = 1{,}981{,}910$.)

```
  D O U B L E
  D O U B L E
      1 O I L +
_____
1 R O U B L E
```

Step 2. Solving alphametic puzzles requires a lot of attention to the numbers that carry over. Each column may include a 1 that's been carried over from the column to the right of it. And the sum in each column may create a 1 that needs to be carried over to the column to the left.

Let's consider the thousands column. We have to add U + U + 1 (together with possibly a carry-over from the sum in the hundreds column), and the answer must equal a number that has U in its units column.

By a process of elimination, the only possible numbers for U are 8 if there is a carried number, since $8 + 8 + 1 + 1 = 18$, or 9 if there is no carried number, since $9 + 9 + 1 = 19$. In both cases a 1 is carried over to the ten-thousands column.

```
  D O U B L E
  D O U B L E
      1 O I L +
  1
_____
1 R O U B L E
```

Step 3. Let's look at the ten-thousands column. We know that O + O + 1 is equal to a number that has O in its units column. The only possible number is O = 9, which means that 1 is carried over to the hundred-thousands column. And since 9 has been taken we now know that U = 8, which means from our calculations above there must also be a carry of 1 in that thousands column.

```
  D 9 8 B L E
  D 9 8 B L E
      1 9 I L +
  1 1 1
_____
1 R 9 8 B L E
```

Step 4. In the hundreds column, the answer has a B in its units position. The two possible sums are either B + B + 9 or, if there's a carried number in that column, B + B + 9 + 1. In the former case B is 1, and in the latter B is 0. Since T is 1, B must be 0, and there is a carried 1 in that column.

```
  D 9 8 0 L E
  D 9 8 0 L E
        1 9 I L +
 1 1 1 1
 ─────────────
1 R 9 8 0 L E
```

Step 5. D must be larger than 5, but it cannot be 5, because that would make R = 1, and that number is already taken. So the only possible values for D and R are D = 6 and R = 3, or D = 7 and R = 5.

Likewise, we can reduce the possibilities for E, L and I.

There are six numbers left that haven't been chosen: 2, 3, 4, 5, 6 and 7.

E cannot be 2, since this would make L = 8, and 8 is taken; nor can E be 5, because in that case L would also have to be 5.

If E = 3, then L = 7, but this combination is impossible with either set of values for D and R.

If E = 7, L = 3, and we have the same problem.

If E = 6, then L = 4 and I = 5, but this combination is again impossible with either set of values for D and R.

But if E = 4, L = 6 and I = 3; therefore D = 7 and R = 5. We're done.

```
  7 9 8 0 6 4
  7 9 8 0 6 4
        1 9 3 6
 1 1 1 1 1
 ─────────────
1 5 9 8 0 6 4
```

(116) ODDS AND EVENS

We're going to solve this long multiplication by separating the two 'short' multiplications within it, which are:

[1] EEO × O = EOEO   and   [2] EEO × O = EOO

Let's start with [2]. The equation states that a three-digit number EEO (the *multiplier*) multiplied by an odd number O (the *multiplicand*) is equal to a three-digit number. The multiplicand cannot be 1, because if it was then the multiplier would be the same as the answer, and it isn't. The multiplier begins with an even number, so it must be at least 201. This tells us that the multiplicand cannot be 5 or higher since 201×5 is 1005, which is a four-digit answer, contradicting the fact that the answer has only three digits. We can deduce therefore that the multiplicand is 3. And if the multiplicand is 3, the first digit of the multiplier must be 2, since if it was 4 or higher, again the answer would have four digits. So, we now have:

[2] $2EO \times 3 = EOO$

The 'tens' digit of the multiplier is even, but the 'tens' digit of the answer is odd. When you multiply an even digit by 3 you get an even digit. So the only way this calculation makes sense is if an odd number is carried over from the multiplication of the 'units' digit of the multiplier by 3. The options for the 'units' digit of the multiplier are now restricted to 5, 7 and 9, since 1 and 3 don't give a carry when multiplied by 3. If the 'units' digit is 5, the number 1 will be carried over (since $5 \times 3 = 15$), and if the 'units' digit is 7 or 9, the number 2 will be carried over (since $7 \times 3 = 21$, and $9 \times 3 = 27$). We know that the carried number must be odd, so the 'units' digit of the multiplier is 5.

[2] $2E5 \times 3 = EOO$

The options for the tens digit of the multiplier are 0, 2, 4, 6 or 8. But we can eliminate 4 and 6 since $245 \times 3 = 735$ and $265 \times 3 = 795$, which contradict the answer (which starts with an even number). So the multiplier is either 205, 225 or 285.

Now let's return to equation [1] with what we know.

[1] One of the following must be true:

  [a] 205 × O = EOEO
  [b] 225 × O = EOEO
  [c] 285 × O = EOEO

The first digit of the answer is even, so the number must be greater than 2000. But when the multiplicand is 1, 3, 5, or 7, the answer in [a], [b] and [c] is less than 2000. So the muliplicand is 9. And when the multiplicand is 9, the only option that satisfies the equation is [c], since the answer in [a] is below 2000 and the answer in [b] is 2025, which contradicts the fact that the second digit must be odd:

[1] 285 × 9 = 2565

So we can now complete the full equation:

$$
\begin{array}{r}
2\,8\,5 \\
\times\ \ 3\,9 \\
\hline
2\,5\,6\,5 \\
8\,5\,5\ \ \\
\hline
1\,1\,1\,1\,5
\end{array}
$$

⑪⑦ THE CROSSWORD THAT COUNTS ITSELF

To prep this puzzle it's worth listing number words according to their length:

Three letters: *one, two, six, ten*
Four letters: *four, five, nine*
Five letters: *three, seven, eight*
Six letters: *eleven, twelve, twenty*
Seven letters: *fifteen, sixteen*
Eight letters: *thirteen, fourteen, eighteen, nineteen*

Step 1. As I told you in the text, 8 Down must be of the form 'ONE *', because all numbers greater than one will include a plural, so will require at least six units.

10 Across is six units, so it must include a three-letter number greater than one, with final letter *. The options are *one, two, six* and *ten*. Yet *one, two* and *ten* would make 8 Down either 'ONE E', 'ONE O' or 'ONE N', which are all contradictions! By elimination, 8 Down is ONE X and 10 Across is 'SIX #s' for some #. (Here and below I'm using different grammatical symbols to stand for different individual letters.)

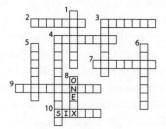

Step 2. 4 Down has an eight-letter number word, so it must be either THIRTEEN, FOURTEEN, EIGHTEEN or NINETEEN Ss. We can eliminate FOURTEEN and NINETEEN because there are no five-letter number words beginning with F or N that we would need to complete 4 Across. We know that there are a total of twelve entries, of which only one is singular (8 Down). The final S in each of the other entries accounts for 11 Ss on the grid. There is an extra S for the SIX at 10 Across, bringing us up to 12 Ss. The only chance for more Ss would be if there are more entries with a SIX, a SEVEN, a SIXTEEN or a SEVENTEEN. But since there are no free entries for three-letter, seven-letter or nine-letter words, we can eliminate SIX, SIXTEEN and SEVENTEEN.

So the extra Ss must come from SEVEN. There are only three possible entries where SEVEN fits, so there are at most 15 Ss. So we can eliminate EIGHTEEN as a candidate for 4 Down, which means that it must be THIRTEEN Ss. Which means that 4 Across must be THREE ?s, and 1 Down must be FOUR @s.

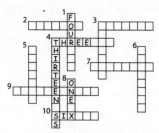

Step 3. If there are thirteen Ss, and twelve are already accounted for, then from our previous calculations there is one entry that contains SEVEN. The only slots

in which SEVEN fits are 6 Down and 7 Across. If it was 7 Across, this would make 3 Down a number of Es. The options are FOUR, FIVE or NINE Es. There are already seven Es on the grid (if we include the SEVEN), so we can eliminate FOUR and FIVE. And we can eliminate NINE, because this would mean that 4 Across is THREE Ns, which is a contradiction, because then there would be four

Ns on the grid (from the NINE, THIRTEEN and SEVEN). So, 6 Down is SEVEN !s.

That means 7 Across is either THREE or EIGHT. We have now used 12 letters: E, F, H, I, N, O, R, S, T, U, V and X. Since we know that there are only 12 letters in the grid, we can eliminate EIGHT, because that contains a G. So, 7 Across is THREE Vs.

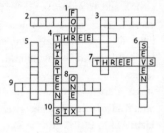

Step 4. 3 Down has to be FOUR Hs, because if it was FIVE then 4 Across would be THREE Vs, but this is the same as 7 Across, which contradicts the fact that there are 12 entries for 12 different letters. Because E is the most common letter of the ones remaining, 9 Across must be THIRTEEN Es. It can't be FOURTEEN, because this would make 5 Down a number of Us, and Us are now counted in 4 Across. And it can't be EIGHTEEN (because there are no Gs) or NINETEEN, because there cannot be as many as nineteen Es in the remaining spaces. The three remaining numbers each have four letters. Since there are THREE Vs but only one V on the grid, two of the remaining entries must be FIVE. And since there are

THREE Us, and two Us on the grid, the final remaining entry must be a FOUR. So, there are four Os in total, which means that 2 Across must be FOUR Os, 5 Down must be FIVE Is, 3 Across must be FIVE Fs and 10 Across must be SIX Ts. And the remaining spaces in 1 Down and 6 Down must be N and R.

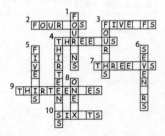

⑪⑧ **AN AUTOBIOGRAPHY IN TEN DIGITS**

I'm going to approach this systematically, starting with solutions that are obviously wrong and then honing in on the answer. The aim is to fill the second row, such that each digit describes how many times the digit above it appears in that row.

Let's imagine the number starts with 9.

0	1	2	3	4	5	6	7	8	9
9									

If this is the case, then the number must have nine 0s, so all the other digits would have to be 0. But we know there is at least one 9, so the other digits cannot all be 0s.

Let's imagine the number starts with 8.

0	1	2	3	4	5	6	7	8	9
8									

This means there are 8 zeros in the nine remaining positions. Since we have at least one 8, the number below it (let's call it x) must be a non-zero digit, which makes all the other slots 0s.

0	1	2	3	4	5	6	7	8	9
8	**0**	**0**	**0**	**0**	**0**	**0**	**0**	**x**	**0**

But there is no value for x that fits! It cannot be 1, since this would contradict the 0 in the second position, which indicates how many 1s there are in the final number. Likewise, all the other possible values for x create contradictions.

Before we continue, we can deduce a certain property of the digits in the second row. They must add up to ten. This is because each number in the row counts how many times a particular digit appears in that row. There are only ten places in the row, so the total digit count must be ten.

Now let's imagine the number starts with 7.

0	1	2	3	4	5	6	7	8	9
7									

We know there are seven zeros. And since there is a 7 in the final number, we know that the digit under 7 in the first row must be non-zero, and either 1, 2 or 3. (If it was larger than 3 then the sum of the digits would be more than 10.) In each case, however, the solution fails. If the digit under 7 is 1, then the digit under 1 would have to be non-zero. This digit could not be 1, because then we'd have two 1s in the final number. The digit under 7 could not be 2, because then the digit under 2 would be non-zero, contradicting the fact that there are seven 0s. And we can use similar logic to show that neither 2 nor 3 will fit under 7.

Next up is 6.

0	1	2	3	4	5	6	7	8	9
6									

There are six zeros. The digit under 6 is non-zero. Let's say it's 1.

0	1	2	3	4	5	6	7	8	9
6						1			

This means that the digit below 1 is non-zero. Since the digits on the bottom row must sum to 10, the digit under 1 could be 1, 2 or 3. We can eliminate 1, because that would mean two 1s in the bottom row, which is a contradiction. If it was 2, then the digit under 2 would have to be 1, so that the digits add up to 10. Things are looking promising! We are left with six empty spaces, which must be the six 0s. And the problem is solved.

0	1	2	3	4	5	6	7	8	9
6	2	1	0	0	0	1	0	0	0

⑲ PANDIGITAL PANDEMONIUM

The number of possible permutations of the ten digits is $10 \times 9 \times 8 \times 7 \times 6 \times 5 \times 4 \times 3 \times 2 \times 1 = 3,628,800$. Each of these permutations is a pandigital number except the ones beginning with 0, since pandigital numbers do not start with a 0. (The permutations that have a 0 in the first position are considered nine-digit numbers.) There are $9 \times 8 \times 7 \times 6 \times 5 \times 4 \times 3 \times 2 \times 1 = 362,880$ of those. So the number of pandigitals is $3,628,800 - 362,880 = 3,265,920$.

⑳ PANDIGITAL AND PANDIVISIBLE

We're going to tick off the digits one by one. Let's start with the low-hanging fruit. Any number divisible by 10 must end in a zero, so $j = 0$. Any number divisible by 5 must end either in a 0 or a 5. So $e = 5$.

With two digits revealed, our number is:

*abcd*5*fghi*0

If a number is divisible by an even number, then that number must itself be even, which means that b, d, f and h must all be even. So b, d, f and h are some combination of 2, 4, 6 and 8. The remaining unknowns, a, c, g and i, are therefore some combination of the remaining (odd) numbers 1, 3, 7 and 9.

Now use the test for divisibility by 4. We know that *abcd* is divisible by 4. So *cd* is also divisible by 4. The only combinations in which c is odd, d is even (which we know from the previous paragraph), and *cd* divides by 4 are 12, 16, 32, 36, 72, 76, 92 and 96. This reveals that d is either 2 or 6.

The test for divisibility by 3 states that if the sum of a number's digits is divisible by 3, then that number is also divisible by 3. It also works the other way around: if a number is divisible by 3, then so is the sum of its digits.

So, $a + b + c$ is divisible by 3.

Anything divisible by 6 is also divisible by 3, so

$a + b + c + d + e + f$ is also divisible by 3.

If two numbers are divisible by 3, then if you subtract the smaller from the larger, the result must also be divisible by 3.

So, the number $a + b + c + d + e + f - (a + b + c) = d + e + f$ is divisible by 3.

We know that d is either 2 or 6; we know that e is 5; and we know that f is either 2, 4, 6 or 8.

If d is 2, then $2 + 5 + f$ must be divisible by 3. So f must be 8. (It cannot be 2, because d is 2 and each digit appears only once. It cannot be 4 or 6, since 11 and 13 are not divisible by 3.)

If d is 6, then $6 + 5 + f$ must be divisible by 3, and by the same reasoning above f must be 4.

So we have two options for the middle three digits. Either def is 258 or it is 654. Let's now try each one.

[1] Option 1: def is 258.

From the divisibility by 8 rule, if an eight-digit number $abcdefgh$ is divisible by 8, then the three-digit number fgh is also divisible by 8.

So, $8gh$ is divisible by 8.

g is 1, 3, 7 or 9 and h is one of the remaining even numbers, 4 or 6. The number $8g4$ is not divisible by 8 for any value of g, so h must be 6. We have accounted for 2, 6 and 8, which means that b, the final even number, must be 4.

So, now our number looks like:

$a4c258g6i0$

Anything divisible by 9 must be divisible by 3. So $a + 4 + c + 2 + 5 + 8 + g + 6 + i$ is divisible by 3. Since anything divisible by 6 is also divisible by 3, we know that $a + 4 + c + 2 + 5 + 8$ is also divisible by 3. As we saw above, if two numbers are divisible by 3, then if you subtract the smaller from the larger, the result must also be divisible by 3. So:

$g + 6 + i$ must be divisible by 3.

So, $g + i$ must be divisible by 3. We have to select g and i from 1, 3, 7 and 9, which means that g and i are each either 3 or 9. So a and c are each either 1 or 7. We therefore have four possibilities for the final number, when a, c, g and i are:

1, 7, 3, 9 (so the number is 1472583690)

7, 1, 3, 9 (so the number is 7412583690)

1, 7, 9, 3 (so the number is 1472589630)

7, 1, 9, 3 (so the number is 7412589630)

Reach for your calculator, and test these numbers to see if they conform to the rules stated in the question. You will find that none of them do.

1472583690 fails because 14725836 is not divisible by 8.

7412583690 fails because 7412583 is not divisible by 7.

1472589630 fails because 1472589 is not divisible by 7.

7412589630 fails because 7412589 is not divisible by 7.

Dead end. This doesn't work. We can conclude that *def* is not 258.

[2] Option 2: *def* is 654.

From the divisibility by 8 rule, if an eight-digit number *abcdefgh* is divisible by 8, then the three-digit number *fgh* is also divisible by 8.

So, 4*gh* is divisible by 8.

Since 4*gh* = 400 + *gh* and 400 is divisible by 8, we know that *gh* is divisible by 8.

g is 1, 3, 7 or 9 and *h* is one of the remaining even numbers, 2 or 8. The number *gh* is not divisible by 8 when *h* is 8, so *h* must be 2. Which means that *b* must be 8.

So, now our number looks like this:

*a*8*c*654*g*2*i*0

Anything divisible by 9 must be divisible by three. So $a + 8 + c + 6 + 5 + 4 + g + 2 + i$ is divisible by 3 and using the reasoning above we know that:

$g + 2 + i$ must be divisible by 3, where *i* and *g* are each one of 1, 3, 7 or 9.

The options for *g* and *i* must be one of these:

1 and 3,

3 and 1,

1 and 9,

9 and 1,

3 and 7,

7 and 3,

7 and 9,

9 and 7.

Let's go through these options and, using a calculator, test the numbers to see if they conform to the rules stated in the question.

1 and 3:

7896541230 – fails because 7896541 is not divisible by 7.
9876541230 – fails because 9876541 is not divisible by 7.

3 and 1:

7896543210 – fails because 7896543 is not divisible by 7.
9876543210 – fails because 9876543 is not divisible by 7.

1 and 9:

7836541290 – fails because 7836541 is not divisible by 7.
3876541290 – fails because 3876541 is not divisible by 7.

9 and 1:

7836549210 – fails because 783654 is not divisible by 6.
3876549210 – fails because 3876549 is not divisible by 7.

3 and 7:

1896543270 – fails because 1896543 is not divisible by 7.
9816543270 – fails because 9816543 is not divisible by 7.

7 and 3:

1896547230 – fails because 1896547 is not divisible by 7.
9816547230 – fails because 9816547 is not divisible by 7.

7 and 9:

1836547290 – fails because 1836547 is not divisible by 7.
3816547290 – **this works.**

9 and 7:

1836549270 – fails because 1836549 is not divisible by 7.
3816549270 – fails because 3816549 is not divisible by 7.

Finally, we found it. The answer is 3816547290.

⑫ 1089 AND ALL THAT

We're trying to find the values for the digits a, b, c and d such that:

$abcd \times 4 = dcba$

The term $abcd$ does not mean $a \times b \times c \times d$; rather, it means that digit a is in the thousands position, b is in the hundreds column, c is in the tens column and d is in the units. So the equation above can be expanded to:

$(1000a + 100b + 10c + d) \times 4 = 1000d + 100c + 10b + a$

The job is now to find clever ways to simplify and solve this equation.

Step 1. Deduce a.

The left-hand side of the equation is even, since it is a multiple of 4, so the right-hand side must also be even. Since $1000d + 100c + 10b$ is also even, we can deduce that a must be even. The left-hand side of the equation puts a limit on a's size, since $4000a$ must be less than 9999. (If it were more than 9999, then the right-hand side would be a five-digit number.) The only even number that fits this condition is 2, so $a = 2$.

Step 2. Deduce d.

Since $a = 2$, the left-hand side of the equation is at least 8000, so d is either 8 or 9. But if $d = 9$, the units digit on the left-hand side would be 6, since $9 \times 4 = 36$. But the units digit has to be 2, because $a = 2$ on the right-hand side. So $d = 8$.

Let's place these values in the original equation:

$(2000 + 100b + 10c + 8) \times 4 = 8000 + 100c + 10b + 2$

Which reduces to:

$8032 + 400b + 40c = 8002 + 100c + 10b$

And again:

$390b + 30 = 60c$

Once more:

$13b + 1 = 2c$

Remember, b and c are single digits. The maximum possible value for c is 9, which means the maximum value for $2c$ is 18. This restricts the value of b to 1. And if $b = 1$, $c = 7$.

The answer is therefore 2178:

$2178 \times 4 = 8712$.

⑫ BACK TO FRONT

We're looking for a number, such that when you double it, the last digit becomes the first digit. Freeman Dyson helpfully told us that the smallest number with this property has 18 digits, so let the number be $nnnnnnnnnnnnnnnnnn_R$, where each n is a digit and n_R is the rightmost digit.

We know that:

$$
\begin{array}{r}
nnnnnnnnnnnnnnnnnn_R \\
\times\,2 \\
\hline
n_R nnnnnnnnnnnnnnnnn
\end{array}
$$

Let's choose a digit for n_R. We can't choose 0, since that would mean the answer only has 17 digits, which makes the multiplication impossible. Nor can we choose 1, because then the answer would start with a 1, which is again impossible: half of an 18-digit number beginning with a 1 is a 17-digit number. But we can choose 2.

So, our multiplication becomes:

$$
\begin{array}{r}
nnnnnnnnnnnnnnnnnn2 \\
\times\,2 \\
\hline
2nnnnnnnnnnnnnnnnn
\end{array}
$$

It is possible to fill in this equation by deducing the values of the ns. Since $2 \times 2 = 4$, we know that the last digit of the bottom number must be 4.

$$
\begin{array}{r}
nnnnnnnnnnnnnnnnnn2 \\
\times\,2 \\
\hline
2nnnnnnnnnnnnnnnn4
\end{array}
$$

The bottom number has the same digits in the same order as the top number, except that the final digit of the top number is the first digit of the bottom number. It follows then that the final digit of the bottom number is the second-last digit of the top number. So the second-last digit of the top number is 4.

$$nnnnnnnnnnnnnnnnn42$$
$$\times 2$$
$$2nnnnnnnnnnnnnnnnn4$$

Since $4 \times 2 = 8$, we know that the second-last digit of the bottom number, and therefore the *third*-last digit of the top number, must be 8.

$$nnnnnnnnnnnnnnnn842$$
$$\times 2$$
$$2nnnnnnnnnnnnnnnn84$$

All the arithmetic we have done so far is the two times table. Our next sum is $8 \times 2 = 16$, so the next slot – third-last number along the bottom, fourth-last number along the top – must be a 6, but since it comes from a 16 we need to hold a 1 for the 'carry'.

$$nnnnnnnnnnnnnnn6842$$
$$\times 2$$
$$2nnnnnnnnnnnnnnn684$$
$$1$$

For the next n we have $6 \times 2 = 12$, but there is also the carried-over 1, making 13. So the next slot – fourth-last number along the bottom, fifth-last number along the top – is 3, again with a carried-over 1.

$$nnnnnnnnnnnnnn36842$$
$$\times 2$$
$$2nnnnnnnnnnnnnn3684$$
$$1$$

If we carry on multiplying out like this, building up both numbers, the process will stop perfectly at:

$$105263157894736842$$
$$\underline{\times 2}$$
$$210526315789473684$$

So, we have an answer: when you double 105263157894736842, the result is the same number with the last digit on the front.

If we had chosen our number to end with a 3, we would have constructed the number 15789473684**2105263**, which doubles to 315789473684210526, and so is also an answer. The digits in bold show how this number contains the same string of digits as the number in the previous paragraph. It is also possible to generate numbers that end in 4, 5, 6, 7, 8 and 9.

(123) THE NINTH POWER

I hope you didn't try to calculate all of them, or, if you did, that you spotted a pattern: the last digit of the ninth power of any number is the last digit of the original number.

So, the numbers in ascending order will end ... 067**1**, ... 883**2**, ... 195**3**, ... 646**4**, ... 187**5**, ... 841**6**, ... 507**7**, ... 284**8**, ... 875**9**.

In fact, here's a table of what happens to the last digit of a number as you head up the powers:

Last digit of n ends in	0	1	2	3	4	5	6	7	8	9
Last digit of n^2 ends in	0	1	4	9	6	5	6	9	4	1
Last digit of n^3 ends in	0	1	8	7	4	5	6	3	2	9
Last digit of n^4 ends in	0	1	6	1	6	5	6	1	6	1
Last digit of n^5 ends in	0	1	2	3	4	5	6	7	8	9
Last digit of n^6 ends in	0	1	4	9	6	5	6	9	4	1
Last digit of n^7 ends in	0	1	8	7	4	5	6	3	2	9
Last digit of n^8 ends in	0	1	6	1	6	5	6	1	6	1
Last digit of n^9 ends in	0	1	2	3	4	5	6	7	8	9

As this table shows, the last digit of the ninth power of any number is the same as the last digit of the original number – but so is the last digit of the fifth power.

In fact, the property of having the same last digit as the original number belongs to the fifth powers, the ninth powers, the thirteenth powers – and so on for every fourth power upwards.

(124) WHEN I'M SIXTY-FOUR

The number 2^{64} is 2 multiplied by itself 64 times. It is also the sixty-fourth term in the doubling sequence, which starts:

2, 4, 8, 16, 32, 64, 128, 256, 512, 1,024...

So far, so familiar. We could carry on for another 54 terms but that would be extremely dull and we'd inevitably make mistakes.

We're looking for something that will help us approximate the answer without having to do too much calculation.

Hang on. The tenth term, or 2^{10}, is 1,024.

That's roughly 1,000, a nice round number.

If a thousand is roughly 2^{10}, then a thousand times itself six times, or $1,000^6$, is roughly 2^{10} times itself six times, which is $(2^{10})^6 = 2^{60}$.

1000^6 = 1,000,000,000,000,000,000, or a quintillion.

So, 2^{60} is roughly a quintillion.

We know that $2^4 = 16$.

So, $2^{64} = 2^{60} \times 2^4$, which is roughly 16 quintillion.

Sixteen quintillion is not a bad estimate, but we can do better, adjusting for our rounding error.

We approximated 1,024 as 1,000. But 1,024 is 2.4 per cent bigger than 1,000. Each time we multiplied a thousand by itself, we should have added an extra 2.4 per cent. Since we multiplied a thousand by itself six times, we should have added 2.4 per cent six times, which is a cumulative increase of about 15 per cent. Our improved answer will be 16 quintillion + 15 per cent.

We can work out 15 per cent of 16 quintillion in our heads. Ten per cent is 1.6

quintillion and five per cent is half that, or 0.8 quintillion. So 15 per cent is an extra 2.4 quintillion.

Our final estimate is (16 + 2.4) quintillion = 18.4 quintillion.

Pretty good, compared with the real answer:

18,446,744,073,709,551,616.

(125) A LOT OF NOTHING

So what *does* it mean to have zeros at the end of a number? This is pretty straightforward. A number ending in 0 is divisible by 10. A number ending in 00 is divisible by 100, which is 10×10. A number ending in 000 is divisible by 1000, which is $10 \times 10 \times 10$. In other words, the zeros at the end of a number denote how many times 10 will divide into that number. The question is therefore asking how many times 100! can be divided by 10.

We know that:

$100! = 100 \times 99 \times 98 \times 97 \times ... \times 3 \times 2 \times 1$.

So let's go through the terms and see which ones divide by 10.

You can divide 10, 20, 30, 40, 50, 60, 70, 80, 90 and 100 by 10, which means there must be at least 11 zeros at the end of 100! (Counting twice for 100, which has two zeros).

But it's not that simple. It is also possible to multiply two numbers that do not end in a 0 to create one that does. Numbers such as:

$8 \times 5 = 40$

$4 \times 15 = 60$

$6 \times 25 = 150$

How can we be sure to detect the zeros in 100! that are created by multiplying numbers that do not end in 0? Let's look more closely at the breakdown of these numbers.

$8 \times 5 = (2 \times 2 \times 2) \times 5$

Which we can rearrange as:

$(2 \times 2) \times (2 \times 5) = 4 \times 10$

Likewise:

$4 \times 15 = (2 \times 2) \times (3 \times 5) = (2 \times 3) \times (2 \times 5) = 6 \times 10$

$6 \times 25 = (3 \times 2) \times (5 \times 5) = (3 \times 5) \times (2 \times 5) = 15 \times 10$

Whenever we multiply two numbers that do not end in a zero to create one that does, there must be a 2 and a 5 in the multiplicative breakdowns of those numbers. This is because whenever a 2 and a 5 appear together in a chain of numbers to be multiplied, you can pair them together to make a 10.

So we can rephrase this puzzle as a search for pairs of 2 and 5 among the multiplicative breakdown of 100!

In fact, we can simplify it even further: essentially, we're looking for the number of times 5 divides into 100!, because 2 obviously divides into 100! many more times than 5 does, so the number of pairs of 2 and 5 is equal to the number of 5s.

How many times does 5 divide into the numbers from 1 to 100? Counting upwards from 1, 5 divides into these numbers:

5, 10, 15, 20, 25 ... 90, 95, 100.

Each of these 20 terms divides by 5 only once, except 25, 50, 75 and 100, which divide twice. So 5 divides into 100! a total of 24 times.

We have our answer. There are 24 zeros at the end of 100!

If you want to check, here's the number in full: 93,326,215,443,944,152,681,699, 238,856,266,700,490,715,968,264,381,621,468,592,963,895,217,599,993,229,915, 608,941,463,976,156,518,286,253,697,920,827,223,758,251,185,210,916,864,000,000, 000,000,000,000,000,000

A List of the Puzzles and a Note on their Sources

The references below include the texts from where I have taken, or adapted, the puzzles in this book. Often these texts are not the original source. In some of the cases, the titles are my own. An asterisk (*) denotes that I have used the original wording of the puzzle (or a translation of the original) in the question.

Every attempt has been made to contact copyright holders. All queries should be addressed to the publisher.

In addition to the books mentioned below I am indebted to the following wonderful resources: David Singmaster's *Sources in Recreational Mathematics*, which although unpublished is easily available online; Alexander Bogomolny's site www.cut-the-knot.org; and St Andrews University's MacTutor History of Mathematics Archive.

Introduction

Number Tree (puzzle p. 1; solution p. 5). Nobuyuki Yoshigahara. *Puzzles 101*, A K Peters/CRC Press (2003).

Canals on Mars (puzzle p. 2; solution p. 6). Sam Loyd. Martin Gardner (ed.), *Mathematical Puzzles of Sam Loyd*, Dover Publications Inc. (2000).

Ten tasty teasers
Are you smarter than an 11-year-old?

(puzzles pp. 7–9; solutions pp. 197–9)

All problems © United Kingdom Mathematics Trust.

Cabbages, Cheating Husbands and a Zebra
LOGIC PROBLEMS

⑩ The Zebra Puzzle (puzzle p. 27; solution p. 207): *Life International* (17 December 1962).

⑪* Caliban's Will (puzzle p. 29; solution p. 210): Hubert Phillips, S. T. Shovelton, G. Struan Marshall, *Caliban's Problem Book*, T. De La Rue (1933).

⑫ Triangular Gunfight (puzzle p. 30; solution p. 211): Hubert Phillips, *Question Time*, J. M. Dent (1937).

⑬ Apples and Oranges (puzzle p. 31; solution p. 212): William Poundstone, *How Would You Move Mount Fuji?*, Little Brown and Co. (2003).

⑭ Salt, Pepper and Relish (puzzle p. 31; solution p. 213): Adapted from Martin Gardner, *My Best Mathematical and Logic Puzzles*, Dover Publications (1994).

⑮ Rock, Paper, Scissors (puzzle p. 32; solution p. 213): Yoshinao Katagiri in Nobuyuki Yoshigahara, *Puzzles 101*, A K Peters/CRC Press (2003).

⑯ Mud Club (puzzle p. 33; solution pp. 33–35): Hubert Phillips, *Week-End*, taken from Hans van Ditmarsch, Barteld Kooi, *One Hundred Prisoners and a Light Bulb*, Springer (2015).

⑰ Soot's You (puzzle p. 36; solution p. 214): George Gamow, Marvin Stern, *Puzzle-Math*, Viking Books (1957).

⑱ Forty Unfaithful Husbands (puzzle p. 37; solution p. 214): George Gamow, Marvin Stern, *Puzzle-Math*, Viking Books (1957).

⑲ Box of Hats (puzzle p. 38; solution p. 216): Kobon Fujimura, *The Tokyo Puzzles*, Biddles Ltd (1978).

⑳ Consecutive Numbers (puzzle p. 40; solution p. 217): Hans van Ditmarsch, Barteld Kooi, *One Hundred Prisoners and a Light Bulb*, Copernicus (2015), based on J. E. Littlewood, *A Mathematician's Miscellany*, Methuen and Co. Ltd (1953).

㉑ Cheryl's Birthday (puzzle p. 41; solution p. 218): Joseph Yeo Boon Wooi, Singapore and Asian Schools Math Olympiad.

㉒ Denise's Birthday (puzzle p. 43; solution p. 219): Joseph Yeo Boon Wooi, theguardian.com.

㉓ The Ages of the Children (puzzle p. 44; solution p. 220): Author unknown.

㉔* Wizards on a Bus (puzzle p. 45; solution p. 221): John Hhorton Conway, Tanya Khovanova, 'Conway's Wizards', *The Mathematical Intelligencer*, vol. 35 (2013).

㉕ Vowel Play (puzzle p. 46; solution p. 225): Peter Wason, 'Wason selection task', Wikipedia.

Ten tasty teasers
Are you a wizard at wordplay?

(puzzles pp. 47–9; solutions p. 227)

Questions 1, 3, 5, 7 and 9 are examples of the game of HIPE, invented by Peter Winkler and featured in his book *Mathematical Mind-Benders*, AK Peters/CRC Press (2007).

Questions 2 and 4 I have seen in many places, but I first read them in Nobuyuki Yoshigahara, *Puzzles 101*, AK Peters/CRC Press (2003).

Question 6: David Singmaster, *Puzzles for Metagrobologists*, World Scientific (2006).

Questions 8 and 10: Author unknown.

A Man Walks Round an Atom
GEOMETRY PROBLEMS

㉖ The Lone Ruler (puzzle p. 53; solution p. 228): The Grabarchuk Family, *The Big, Big, Big Book of Brainteasers*, Puzzlewright (2011).

㉗ Rope Around the Earth (puzzle p. 55; solution p. 229): Author unknown.

(28) Bunting for the Street Party (puzzle p. 58; solution p. 230): Based on a conversation with Colin Wright.

(29) On Yer Bike, Sherlock! (puzzle p. 60; solution p. 231): Joseph D. E. Konhauser, Dan Velleman, Stan Wagon, *Which Way Did the Bicycle Go?*, The Mathematical Association of America (1997).

(30) Fuzzy Math (puzzle p. 62; solution p. 232): Based on an idea in Joseph D. E. Konhauser, Dan Velleman, Stan Wagon, *Which Way Did the Bicycle Go?*, The Mathematical Association of America (1997).

(31) Round in Circles (puzzle p. 63; solution p. 232): *New York Times* (25 May 1982).

(32) Eight Neat Sheets (puzzle p. 64; solution p. 233): Kobon Fujimura, *The Tokyo Puzzles*, Biddles Ltd (1978).

(33) A Square of Two Halves (puzzle p. 65; solution p. 234): Kobon Fujimura, *The Tokyo Puzzles*, Biddles Ltd (1978).

(34) The Wing and the Lens (puzzle p. 66; solution p. 234): Kobon Fujimura, *The Tokyo Puzzles*, Biddles Ltd (1978).

(35) Sangaku Circles (puzzle p. 68; solution p. 235): H. Fukagawa, A. Rothman, *Sacred Mathematics: Japanese Temple Geometry*, Princeton University Press (2008).

(36) Sangaku Triangle (puzzle p. 69; solution p. 237): H. Fukagawa, A. Rothman, *Sacred Mathematics: Japanese Temple Geometry*, Princeton University Press (2008).

(37) Treading on the Tatami (puzzle p. 70; solution p. 238): Kobon Fujimura, *The Tokyo Puzzles*, Biddles Ltd (1978).

(38) Fifteen Tatami Mats (puzzle p. 71; solution p. 238): Donald Knuth, *The Art of Computer Programming*, Addison-Wesley (1968).

(39) Nob's Mats (puzzle p. 72; solution p. 239): Nobuyuki Yoshigahara, *Puzzles 101*, A K Peters/CRC Press (2003).

Ten tasty teasers
Are you smarter than a 12-year-old?

Mathematics Most Fowl
PRACTICAL PROBLEMS

(51) One Hundred Fowls (puzzle p. 98; solution p. 100): David Singmaster, *Sources in Recreational Mathematics*.

(52) One Hundred Birds (puzzle p. 101; solution p. 249): Abu Kamil, *Book of Birds* (n.d.).

(53) The 7-Eleven (puzzle p. 102; solution p. 250). Author unknown.

(54) The Three Jugs (puzzle p. 104; solution p. 252): Abbott Albert, *Annales Stadenses* (13th century).

(55) The Two Buckets (puzzle p. 107; solution p. 252): Adaptation of the Three Jugs puzzle.

(56) The White Coffee Problem (puzzle p. 108; solution p. 252): Author unknown.

(57) Water and Wine (puzzle p. 108; solution p. 253): Martin Gardner, *My Best Mathematical and Logic Puzzles*, Dover Publications (1994).

(58) Famous for 15 Minutes (puzzle p. 109; solution p. 254): Yuri B. Chernyak, Robert M. Rose, *The Chicken from Minsk*, Basic Books (1995).

(59) A Fuse to Confuse (puzzle p. 110; solution p. 255): (i) Author unknown. (ii) 'Time to Burn', Varsity Math week 25, *Wall Street Journal*; and MoMath.org.

(60) The Biased Coin (puzzle p. 111; solution p. 256): Attributed to John von Neumann.

(61) Divide the Flour (puzzle p. 111; solution p. 256): Adapted from Boris A. Kordemsky, *The Moscow Puzzles*, Dover Publications (1955).

(62) Bachet's Weight Problem (puzzle p. 113; solution p. 256): Claude-Gaspard Bachet, *Problèmes Plaisants & Délectables Qui Se Font Par Les Nombres*, 5th edn, A. Blanchard (1993).

(63) The Counterfeit Coin (puzzle p. 115; solution p. 258): Boris A. Kordemsky, *The Moscow Puzzles*, Dover Publications (1955).

(64) The Fake Stack (puzzle p. 116; solution p. 260): Martin Gardner, *My Best Mathematical and Logic Puzzles*, Dover Publications (1994).

(65) From Le Havre to New York (puzzle p. 117; solution p. 260): Charles-Ange Laisant, *Initiation mathématique*, Hachette (1915).

(66) The Round Trip (puzzle p. 118; solution p. 261): William Poundstone, *Are You Smart Enough to Work at Google?*, Little, Brown and Co. (2012).

(67) The Mileage Problem (puzzle p. 119; solution p. 262): Harry Nelson in Scott Kim, *The Little Book of Big Mind Benders*, Workman Publishing (2014).

(68) The Overtake (puzzle p. 119; solution p. 263): Dick Hess, *Mental Gymnastics*, Dover Publications (2011).

(69) The Running Styles (puzzle p. 120; solution p. 263): Joseph D. E. Konhauser, Dan Velleman, Stan Wagon, *Which Way Did the Bicycle Go?*, The Mathematical Association of America (1997).

(70) The Shrivelled Spuds (puzzle p. 120; solution p. 265): Author unknown.

(71) The Wage Wager (puzzle p. 122; solution p. 266): W. W. Rouse Ball, *Mathematical Recreations and Essays*, Project Gutenberg (1892).

(72) A Sticky Problem (puzzle p. 122; solution p. 267): Frederick Mosteller, *Fifty Challenging Problems in Probability*, Dover Publications (1965).

(73) The Handshakes (puzzle p. 123; solution p. 267): Author unknown.

(74) The Handshakes and the Kisses (puzzle p. 123; solution p. 268): The Grabarchuk Family, *The Big, Big, Big Book of Brainteasers*, Puzzlewright (2011).

(75) The Lost Ticket (puzzle p. 124; solution p. 269): Peter Winkler, *Mathematical Puzzles*, A K Peters/CRC Press (2003).

Ten tasty teasers
Are you a genius at geography?

(puzzles pp. 125–6; solutions p. 272)

The idea of having geography questions in a book of mathematical puzzles is borrowed from Peter Winkler, who did the same in *Mathematical Puzzles*, A K Peters/CRC Press (2003). Some of my questions are inspired by his, and they all involve some kind of mathematical thinking.

Your Aid I Want, Nine Trees to Plant
PROBLEMS WITH PROPS

Four Coins (puzzle p. 128; solution p. 129): H. E. Dudeney, *536 Puzzles & Curious Problems*, Scribner Book Co. (1983).

76 The Six Coins (puzzle p. 129; solution p. 273): H. E. Dudeney, *536 Puzzles & Curious Problems*, Scribner Book Co. (1983).

77 Triangle to Line (puzzle p. 130; solution p. 273): Erik Demaine, Martin Demaine, 'Sliding Coin Puzzles' in *Tribute to a Mathemagician*, A K Peters/ CRC Press (2004).

78 The Water Puzzle (puzzle p. 131; solution p. 274): Nobuyuki Yoshigahara, *Puzzles 101*, A K Peters/CRC Press (2003), and Erik Demaine and Martin Demaine, 'Sliding Coin Puzzles' in *Tribute to a Mathemagician*, A K Peters/ CRC Press (2004).

79* The Five Pennies (puzzle p. 132; solution p. 274): H. E. Dudeney, *Amusements in Mathematics*, Project Gutenberg (1958), and Kobon Fujimura, *The Tokyo Puzzles*, Biddles Ltd (1978).

80 Planting Ten Trees (puzzle p. 134; solution p. 275): H. E. Dudeney, *Amusements in Mathematics*, Project Gutenberg (1958).

81 The Space Race (puzzle p. 137; solution p. 276): H. E. Dudeney, *Amusements in Mathematics*, Project Gutenberg (1958).

(82) Tait's Teaser (puzzle p. 138; solution p. 277): P. G. Tait, Introductory Address to the Edinburgh Mathematical Society, Nov 9, 1883, found in *Philosophical Magazine* (January 1884). Extra puzzle: Martin Gardner, *My Best Mathematical and Logic Puzzles*, Dover Publications (1994).

(83) The Four Stacks (puzzle p. 140; solution p. 278): Edouard Lucas, *Recreations Mathematiques*.

(84) Frogs and Toads (puzzle p. 141; solution p. 278): Edouard Lucas, *Recreations Mathematiques*.

(85) Triangle Solitaire (puzzle p. 142; solution p. 279): Martin Gardner, *Mathematical Carnival*, Alfred A. Knopf (1975).

(86) Coins in the Dark (puzzle p. 144; solution p. 279): Author unknown.

(87) The One Hundred Coins (puzzle p. 145; solution p. 280): Gyula Horváth, International Olympiad in Informatics 1996, in Peter Winkler, *Mathematical Puzzles*, A K Peters/CRC Press (2003).

(88) Free the Coin (puzzle p. 146; solution p. 281): Jack Botermans, *Matchstick Puzzles*, Sterling (2007).

(89) Pruning Triangles (puzzle p. 146; solution p. 281): H. E. Dudeney, *536 Puzzles & Curious Problems*, Scribner Book Co. (1983).

(90) Triangle, and Triangle Again (puzzle p. 147; solution p. 281): Kobon Fujimura, *The Tokyo Puzzles*, Biddles Ltd (1978).

(91) Growing Triangles (puzzle p. 148; solution p. 282): (i) The Grabarchuk Family, *The Big, Big, Big Book of Brainteasers*, Puzzlewright (2011). (ii) Author unknown.

(92) A Touching Problem (puzzle p. 149; solution p. 282): Martin Gardner, *My Best Mathematical and Logic Puzzles*, Dover Publications (1994).

(93) Point to Point (puzzle p. 149; solution p. 282): Joseph D. E. Konhauser, Dan Velleman, Stan Wagon, *Which Way Did the Bicycle Go?*, The Mathematical Association of America (1997).

94 The Two Enclosures (puzzle p. 150; solution p. 283): H. E. Dudeney,
536 Puzzles & Curious Problems, Scribner Book Co. (1983).

95 Folding Stamps (puzzle p. 151; solution p. 283): H. E. Dudeney, *536 Puzzles
& Curious Problems*, Scribner Book Co. (1983).

96 The Four Stamps (puzzle p. 152; solution p. 284): H. E. Dudeney, *Amusements
in Mathematics*, Project Gutenberg (1958).

97 The Broken Chessboard (puzzle p. 154; solution p. 285): H. E. Dudeney,
The Canterbury Puzzles, E. P. Dutton and Co. (1908).

98 Folding a Cube (puzzle p. 155; solution p. 285): Nobuyuki Yoshigahara,
Puzzles 101, A K Peters/CRC Press (2003).

99 The Impossible Braid (puzzle p. 156; solution p. 286): Author unknown.

100 Tangloids (puzzle p. 158): Martin Gardner, *New Mathematical Diversions*,
The Mathematical Association of America (1996).

Ten tasty teasers
Are you smarter than a 13-year-old?

(puzzles pp. 161–3; solutions pp. 280–290)

All problems © United Kingdom Mathematics Trust.

The Number Games
PROBLEMS FOR PURISTS

101 Mirror, Mirror (puzzle p. 168; solution p. 291): Boris A. Kordemsky,
The Moscow Puzzles, Dover Publications (1955).

102 Nous Like Gauss (puzzle p. 168; solution p. 291): Derrick Niederman,
Math Puzzles for the Clever Mind, Sterling (2001).

⑩③ That's Sum Table (puzzle p. 169; solution p. 292): Anany Levitin, Maria Levitin, *Algorithmic Puzzles*, Oxford University Press (2011).

⑩④ The Square Digits (puzzle p. 170; solution p. 293): Kobon Fujimura, *The Tokyo Puzzles*, Biddles Ltd (1978).

⑩⑤ The Ghost Equations (puzzle p. 170; solution p. 293): Nobuyuki Yoshigahara, *Puzzles 101*, A K Peters/CRC Press (2003).

⑩⑥ Ring my Number (puzzle p. 170; solution p. 294): Nobuyuki Yoshigahara, *Puzzles 101*, A K Peters/CRC Press (2003).

⑩⑦ The Four Fours (puzzle p. 172; solution p. 294): Solutions thanks to http://mathforum.org/ruth/four4s.puzzle.html.

⑩⑧ Our Columbus Problem (puzzle p. 174; solution p. 295): Sam Loyd in Martin Gardner (ed.), *More Mathematical Puzzles of Sam Loyd*, Dover Publications (1960).

⑩⑨ Threes and Eights (puzzle p. 175; solution p. 296): Author unknown.

⑪⑩ Child's Play (puzzle p. 176; solution p. 296): Author unknown.

⑪⑪ Follow the Arrow 1 (puzzle p. 176; solution p. 297): Nobuyuki Yoshigahara, *Puzzles 101*, A K Peters/CRC Press (2003).

⑪⑫ Follow the Arrow 2 (puzzle p. 177; solution p. 297): Nobuyuki Yoshigahara, *Puzzles 101*, A K Peters/CRC Press (2003).

⑪⑬ Follow the Arrow 3 (puzzle p. 177; solution p. 298): William Poundstone, *Are You Smart Enough to Work at Google?*, Little, Brown and Co. (2012).

⑪⑭ Dictionary Corner (puzzle p. 178; solution p. 299): Dick Hess, *Mental Gymnastics*, Dover Publications (2011).

⑪⑮ The Three Witches (puzzle p. 182; solution p. 302): Mike Keith, http://www.cadaeic.net/alphas.htm.

⑪⑯ Odds and Evens (puzzle p. 183; solution p. 303): Martin Gardner, *The Unexpected Hanging and Other Mathematical Diversions*, University of Chicago Press (1963).

(117) The Crossword that Counts Itself (puzzle p. 184; solution p. 305): Lee Sallows in Joseph D. E. Konhauser, Dan Velleman, Stan Wagon, *Which Way Did the Bicycle Go?*, The Mathematical Association of America (1997).

(118) An Autobiography in Ten Digits (puzzle p. 187; solution p. 308): Martin Gardner, *Mathematical Circus*, Vintage Books (1968).

(119) Pandigital Pandemonium (puzzle p. 188; solution p. 310): Ivan Moscovich, *The Big Book of Brain Games*, Workman Publishing (2006).

(120) Pandigital and Pandivisible? (puzzle p. 189; solution p. 310): Author unknown.

(121) 1089 and All That (puzzle p. 190; solution p. 314): Joseph D. E. Konhauser, Dan Velleman, Stan Wagon, *Which Way Did the Bicycle Go?*, The Mathematical Association of America (1997).

(122) Back to Front (puzzle p. 191; solution p. 315): *New York Times* online (6 April 2009).

(123) The Ninth Power (puzzle p. 192; solution p. 317): Derrick Niederman, *Math Puzzles for the Clever Mind*, Sterling (2001).

(124) When I'm Sixty-four (puzzle p. 192; solution p. 318): William Poundstone, *Are You Smart Enough to Work at Google?*, Little, Brown and Co. (2012).

(125) A Lot of Nothing (puzzle p. 193; solution p. 319): William Poundstone, *Are You Smart Enough to Work at Google?*, Little, Brown and Co. (2012).

Acknowledgements

I am hugely grateful to the United Kingdom Mathematics Trust its director Rachel Greenhalgh for allowing me to reprint questions from its Junior Challenges in the Ten Tasty Teasers sections. The UKMT is a wonderful charity that inspires hundreds of thousands of children to enjoy maths by organising national competitions, as well as training our best pupils to compete in the International Mathematical Olympiad. About 300,000 British children aged 11 to 13 take the Junior Challenge every year. For more information, and resources, the website is www.ukmt.org.uk.

Thanks also to the many mathematicians, puzzle experts, enthusiasts and inventors who I have spoken to writing my blog and researching this book, especially David Singmaster, Tanya Khovanova, Joseph Yeo Boon Wooi, Colin Wright, Naoki Inaba, Maki Kaji, Jimmy Goto, Dick Hess, Gary Foshee, John Conway, Colin Beveridge, Adam P. Goucher, Hans van Ditmarsch, Lee Sallows, Mike Keith, Bill Ritchie, Richard Forster, Otto Janko, James Marshall and Alejandro Erickson.

Laura Hassan has been a brilliant and inspiring editor. Lindsay Davies has masterfully and heroically steered the book through production. Rich Carr's elegant design, helped by Ruth Murray and Ruth Rudd, has made this the loveliest-looking puzzle book I've seen. Thanks also to Andrew Joyce for his fantastic cartoons. Ben Sumner is the smartest copy editor in the business, and Hamish Ironside the most mathematically astute proof reader. At Faber I'm also indebted to Lauren Nicoll in publicity, Lindsay Terrell in marketing, Jack

Murphy in production, Alex Kirby in design and Sara Montgomery at Guardian Faber.

A shout out also to my agent Rebecca Carter and her colleagues Emma Parry, Rebecca Folland and Kirsty Gordon at Janklow & Nesbit.

Respect to the *Guardian* for backing my puzzle blog in the first place, and to the support there from Tash Reith-Banks, Ian Anderson, Pete Etchells and James Randerson.

These friends all contributed in some way too; I'll leave it for them to puzzle out how: Edmund Harriss, Siobhan Roberts, Matt McAllester, Chieko Tsuneoka, Sam Cartmell and Annette Mackenzie.

Lastly, I could not have written this book without the love and support of my wife Natalie. The fact that I managed to write this book during the first year of our son Zak's life is as much down to her as it is to me.